Management and Industrial Engineering

Series Editor

J. Paulo Davim, Department of Mechanical Engineering, University of Aveiro, Aveiro, Portugal

This series fosters information exchange and discussion on management and industrial engineering and related aspects, namely global management, organizational development and change, strategic management, lean production, performance management, production management, quality engineering, maintenance management, productivity improvement, materials management, human resource management, workforce behavior, innovation and change, technological and organizational flexibility, self-directed work teams, knowledge management, organizational learning, learning organizations, entrepreneurship, sustainable management, etc. The series provides discussion and the exchange of information on principles, strategies, models, techniques, methodologies and applications of management and industrial engineering in the field of the different types of organizational activities. It aims to communicate the latest developments and thinking in what concerns the latest research activity relating to new organizational challenges and changes world-wide. Contributions to this book series are welcome on all subjects related with management and industrial engineering. To submit a proposal or request further information, please contact Professor J. Paulo Davim, Book Series Editor, pdavim@ua.pt

More information about this series at https://link.springer.com/bookseries/11690

Panagiotis Kyratsis · Nikolaos Efkolidis ·
J. Paulo Davim
Editors

Advances in Product Design Engineering

Springer

Editors
Panagiotis Kyratsis
Department of Product and Systems Design Engineering
University of Western Macedonia
Kila Kozani, Greece

Nikolaos Efkolidis
Department of Product and Systems Design Engineering
University of Western Macedonia
Kila Kozani, Greece

J. Paulo Davim
Department of Mechanical Engineering
University of Aveiro
Aveiro, Portugal

ISSN 2365-0532 ISSN 2365-0540 (electronic)
Management and Industrial Engineering
ISBN 978-3-030-98123-5 ISBN 978-3-030-98124-2 (eBook)
https://doi.org/10.1007/978-3-030-98124-2

This Springer imprint is published by the registered company Springer Nature Switzerland AG
The registered company address is: Gewerbestrasse 11, 6330 Cham, Switzerland

Preface

The present book brings together a variety of topics that deal with the product design process. It aims in following a holistic approach and incorporates a number of directions, i.e. conceptual design, design methodologies, sustainability, manufacturing, product analysis, materials, design and manufacturing technologies. Product design by itself is a very trendy topic, especially nowadays that the product lifecycle has been decreased and the available technologies become more advanced. It involves processes with increased value and can be extended to areas that computerized approaches are not very common.

A number of high-quality works have been incorporated within the following chapters. The initial Chapter "Industrial Design: Shaped by Technology and Users. Past, Present, AM and the Future" examines how industrial design has influenced and been influenced by additive manufacturing, viewed in the context of the major technological breakthroughs of the last century and the different views of the user brought about by societal change, a key factor for industrial designers. The Chapter "Using the Axiomatic Design in Engineering" aims in presenting the methodology of Axiomatic Design (AD) that can be applied in order to solve a wide variety of problems. It is possible to apply AD to subjects in humanities and social sciences. Properly using those axioms leads to the best design solution for products, processes or systems. The Chapter "4D Printing on Textiles: Developing a File to Fabrication Framework for Self-Forming, Composite Wearables" presents and compares two methods of "file to fabrication" techniques for generating self-forming textile shell structures: The first is based on printed patterns related to their stress line simulation and the second on modified geometrical patterns in relation to their curvature analysis. Furthermore, the buckling degree of the composites in relation to their fabric thickness and elasticity is investigated. The Chapter "Personalized Fashion On-Demand and e-Fashion Business Models: A User Survey in Greece" aims in identifying the reasons for which fashion on demand in Greece is only seen on a restricted scale, by investigating consumer attitudes towards e-commerce and their predispositions towards a new platform of personalized products. Chapter "Machine Learning to Classify and Predict Design and Fabrication Solutions of Architectural Prototypes Driven by Sustainable Criteria" deals with a methodology for using

Machine Learning (ML) in an attempt to classify and predict design solutions of complex structures and products, considering static and geometric criteria as well as criteria of sustainable fabrication during the development of small-scale prototypes. In order to achieve this, a series of approaches are incorporated in the suggested methodology, including Topology Optimization (TO) analysis, design solutions development, post-processing analysis of the design solutions and ML investigation. Chapter "Evaluation of the Station Uniform for Firefighters by Anthropometric Method on the Basis of the Physico-mechanical Properties of the Material" aims in testing firefighters' clothing in different body positions during their work, using the anthropometric method developed at TTK University of Applied Sciences, according to which the suitability of the size number and height of a firefighter's station uniform can be assessed. Research conducted with the Rescue Board revealed that more complex body positions test the properties of special clothing materials and cause discomfort to the wearer. Chapter "Development of Modified Polymethyl Methacrylate and Hydroxyapatite (PMMA/HA) Biomaterial composite for Orthopaedic Products" explores the feasibility of PMMA by the supplement of Hydroxyapatite (HA) based bone cement at varying ranges of 4–16 wt.%. The supplement of 10% HA nanoparticles to the powder of PMMA bone cement increased flexural strength by 12.8% and compression strength by 21.2%. The distribution of reinforced nanoparticles in the nanocomposite was studied using Fourier Transform Infrared (FTIR) spectroscopy and X-Ray Diffraction (XRD). With the help of Scanning Electron Microscopy (SEM) study of the fractured surface of the proposed nanocomposite bone cement, the reason behind the change in mechanical properties was observed. Chapter "Conceptual Design for Innovation: Process and a Knowledge-Based Approach" introduces the crucial initial stage of product innovation design. It focuses on method to innovation design and knowledge-based approach that leads the user through the design process, helps identify critical issues and propose new conceptual schemes for specific configurations. The definition and types of innovative conceptual design are introduced and the factors that affect individual creativity are analysed. A process of innovative conceptual design is proposed. Based on the characteristics of product design, cognitive psychology, knowledge-based approach and information technology is incorporated into design technology. Based on the cognising process, the design activities are defined, knowledge is recommended, and then, suitable creative solutions are evaluated and selected. Chapter "Establishing Product Appearance Specifications with the Identification of User Aesthetic Needs in Product Conceptual Design" presents the basic concepts of aesthetic needs and product specifications in conceptual design. The user aesthetics needs are considered one of the significant determinants in increasing user satisfaction. In this regard, the importance of establishing product appearance specifications to identify user aesthetic needs is discussed. A method is introduced to demonstrate the significance of considering aesthetic and emotional needs when establishing product appearance specifications in product conceptual design. The final Chapter "Digital Transformation of the Product Design and Idea Generation Process" deals with the transformation of the product design and idea generation process influenced by modern digital technologies. Influences on the first stages of the product design process are considered. Market research as the information source is subject to drastic changes due to the introduction of data gathering techniques in the digital era and emerging of the product intelligence process and tools. The introduction of artificial intelligence in concept design and idea generation offers exciting possibilities in creative design problem solving, suggesting future automation of the design process.

The editors acknowledge the aid of Springer Publications and express their gratitude for this opportunity and for their professional support. The editors also express their gratitude to all the chapter authors for their availability and for delivering their high-quality research work.

Kila Kozani, Greece — Panagiotis Kyratsis
Kila Kozani, Greece — Nikolaos Efkolidis
Aveiro, Portugal — J. Paulo Davim
February 2022

Contents

About the Editors

Prof. Panagiotis Kyratsis (www.kyratsis.com) is a Professor in the Department of Product and Systems Design Engineering, University of Western Macedonia, Greece. He is the Director of the Computational Design and Digital Fabrication Research Lab and the Director of the Institute of Traditional Architecture and Cultural Heritage, University Research Center "TEMENUS" (urc.uowm.gr). Professor Panagiotis Kyratsis received his Ph.D. in the area of CAD-based manufacturing process simulations from the Department of Production Engineering and Management, Technical University of Crete, Greece. He holds a diploma in Mechanical Engineering from the Aristotle's University of Thessaloniki—Greece, and he received his M.Sc. in Automotive Product Engineering and M.Sc. in CAD/CAM from Cranfield University—UK, in 1997 and 1999, respectively. He has been involved in a number of industrial projects and he has a great deal of expertise in both the design and the manufacturing aspects of product development. His main research interests include manufacturing, machining, CAD/CAM/CAE systems, product design, reverse engineering and prototyping. He has published more than 20 books and more than 150 papers in Scientific Journals and International Conferences. He acts as member of the editorial board and reviewer to numerous scientific journals and holds 12 industrial designs and one patent registered within the Greek Patent Office.

Dr. Nikolaos Efkolidis (www.efkolidis.eu) is an Assistant Professor in the Department of Product and Systems Design Engineering, University of Western Macedonia, Greece. He received his Ph.D. in the area of Sustainability "Development of Sustainable Methodologies in Product Design, Manufacturing and Education", from the University of Zaragoza, Spain. He holds a diploma in Industrial Design from the University of Western Macedonia, Greece and an M.Sc. in Product Design and Management, from the University of Liverpool, UK. His main research interests include product design, sustainable design, manufacturing, machining, CAD/CAM/CAE systems, reverse engineering and prototyping. He is a co-founder of the Computational Design and Digital fabrication Lab (CODE+) and heptahedron.studio. He has published 5 books and more than 35 papers in Scientific Journals and International Conferences. He acts as a reviewer to a scientific journal and holds 5 industrial designs registered within the Greek Patent Office.

Prof. J. Paulo Davim is a Full Professor at the University of Aveiro, Portugal. He is also distinguished as honorary professor in several universities/colleges/institutes in China, India and Spain. He received his Ph.D. degree in Mechanical Engineering in 1997, M.Sc. degree in Mechanical Engineering (materials and manufacturing processes) in 1991, Mechanical Engineering degree (5 years) in 1986, from the University of Porto (FEUP), the Aggregate title (Full Habilitation) from the University of Coimbra in 2005 and the D.Sc. (Higher Doctorate) from London Metropolitan University in 2013. He is Senior Chartered Engineer by the Portuguese Institution of Engineers with an MBA and Specialist titles in Engineering and Industrial Management as well as in Metrology. He is also Eur Ing by FEANI-Brussels and Fellow (FIET) of IET-London. He has more than 30 years of teaching and research experience in Manufacturing, Materials, Mechanical and Industrial Engineering, with special emphasis in Machining & Tribology. He has also interest in Management, Engineering Education and Higher Education for Sustainability. He has guided large numbers of postdoc, Ph.D. and master's students as well as has coordinated and participated in several financed research projects. He has received several scientific awards and honours. He has worked as evaluator of projects for ERC-European Research Council and other international research agencies as well as examiner of Ph.D. thesis for many universities in different countries. He is the Editor in Chief of several international journals, Guest Editor of journals, books Editor, book Series Editor and Scientific Advisory for many international journals and conferences. Presently, he is an Editorial Board member of 30 international journals and acts as reviewer for more than 100 prestigious Web of Science journals. In addition, he has also published as editor (and co-editor) more than 200 books and as author (and co-author) more than 15 books, 100 book chapters and 500 articles in journals and conferences (more than 280 articles in journals indexed in Web of Science core collection/h-index 59+/11500+ citations, SCOPUS/h-index 63+/14000+ citations, Google Scholar/h-index 82+/23500+ citations). He has been listed in World's Top 2% Scientists by Stanford University study.

Industrial Design: Shaped by Technology and Users. Past, Present, AM, and the Future

Steinar W. Killi

This chapter examines how industrial design has influenced and been influenced by additive manufacturing, viewed in the context of the major technological breakthroughs of the last century, and the different views of the user brought about by societal change, a key factor for industrial designers. A holistic method for designing for additive manufacturing is presented, and emerging technologies such as virtual and augmented reality are introduced as a future game changer for industrial design and user involvement.

1 The Past

The industrial revolution did not just occur in one period [1], but in three. There is also a notion that we are today entering the fourth period of industrial revolution [2]. Whether the industrial revolution is made up of two, three or more periods is a subject of discussion [3, 4]. So too is when the periods started and how long they lasted. I will, however, use the three industrial revolution model in the following [4], toward the end of the chapter discussing whether additive manufacturing is part of a fourth industrial revolution. The industrial revolutions are considered to have had a significant impact on and are therefore used as a backdrop to industrial design, later additive manufacturing possibly being a key facilitator of the fourth industrial revolution [5, 6].

S. W. Killi (✉)
Oslo School of Architecture and Design, Oslo, Norway
e-mail: Steinar.killi@aho.no

P. Kyratsis et al. (eds.), *Advances in Product Design Engineering*, Management and Industrial Engineering,
https://doi.org/10.1007/978-3-030-98124-2_1

The first industrial revolution period started around 1760, and was driven by the introduction of steam power, textile factories, and labor saving production methods, the period fading around 1860 [7]. This first industrial revolution was started and driven by technology, changed the lives of everyone, in the western world, and spurred the new capitalism [8]. It was also a prerequisite for industrial design, as it introduced the concept of mass production. Changes in this period were, in the first period, primarily in raw materials, the introduction of loom produced textiles, and the industrialization of farming. The second industrial revolution was therefore the period that gave birth to the industrial designer, the period starting in around 1870 and fading with the outbreak of World War One. The changes in this second industrial revolution period [9] included milestones such as standardization, manufacturing methods, machine tools, and interchangeable parts, this period therefore representing one in which end product or consumer product production was industrialized, a prerequisite for the emergence of the industrial designer. Products for large-scale or mass production therefore needed to be designed, the industrial designer emerging to fill that role. The third and the possible coming fourth industrial revolution [10] has changed the role of the industrial designer, the second industrial revolution giving birth and defining the industrial designer, which I will describe below.

Industrial design was brought about by developments in technology [11], the birth of the profession being a direct result of the invention and adoption of mass production and the assembly line [12]. This not only changed production methods, but also design itself, the design process morphing from designing for the few to designing for the many. Two professions, the artisan and the engineer, were brought together by this change and fused into one, there being good grounds to claim that the industrial designer is the child of these two professions [13–15]. Industrialization of design furthermore brought the emergence of ergonomics [16], the first phase being measure and fittings, later transitioning into user needs. Henry Dreyfuss' studies in anthropometry [17], and actual measurements of the human body, also paved the way for a closer relation to the user, their needs, and limitations. It was, however, as is so often the case a war, the second world war, that led to the enhanced impact of ergonomics, primarily in the positioning of knobs and controls on airplanes [18]. The involvement of users would become increasingly important in the following years, the understanding and interaction with users being one of the most important skills of an industrial designer [19]. I will return to this throughout this chapter.

Institutions that educate industrial designers usually have one foot in both the artisan and engineer professions, often more in one than the other, many polytechnical schools offering bachelor's and master's degrees in both engineering and industrial design. Art schools also offer bachelor's and master's degrees in both arts and crafts and industrial design [20]. Industrial designers are therefore a diverse group, spanning from artists who industrialize their crafts using mass production methods, to engineers who focus on anthropometry [21] and production ability. The most famous representative of the latter is probably Henry Ford, whose concept of designing for manufacturability [22, 23] was a paradigm change in product design, his only problem being said to be how to produce enough cars. There are also good

Fig. 1 The Wassily Chair, an iconic design by Marcel Breuer, depicted on a German stamp in 1998 (Photo, Shutterstock)

examples of artists who transformed their crafts to achieve simpler and faster production. Bauhaus designer Marcel Breuer designed his iconic Wassily chair, see Fig. 1, after being inspired by his Adler bicycle frame of seamless steel tubing, developed by Mannesmann, which was bent into shape without breaking [24]. The Wassily chair changed furniture design. It was first built by the artist/designer, its complex expression combined with simple construction making it a "must have design object" even today 100 years later.

The role of industrial design is, as stated above, to design products for the many. I will, therefore, in the following, highlight some of the many synergies and paradigm changes that have taken place in industrial design over the years.

An American Chemist, Earl Tupper, in 1946 introduced a new product, a plastic food storage container. His invention was not the box, but the durable plastic it was made of, PE, he designed the box using the material he had invented [25]. It was at first not a success, collecting dust on store shelves, but became an iconic marketing story. Housewife Brownie Wise discovered the product and started to sell it to other housewives at home parties, which are often called Tupperware parties. These parties do still exist, and are still the only way to get hold of the genuine item, see Fig. 2. Plastic containers made of polyethylene for food storage are, of course, today ubiquitous and produced by a plethora of manufacturers. The success of Tupperware, however, began with the invention of a material, and was brought to fruition by a marketing process driven by user insight.

Tupperware parties were not only a new way of selling products, through social gatherings, but also a source of input from users. The user groups, the housewives, were quite homogeneous, but were also an expert group that could give feedback on current products and on potential new products. It's not quite co-creating, but the

Fig. 2 Tupperware products (Photo Mona Svendsen)

home parties definitely provided good insight and valuable feedback to designers [26].

The third industrial revolution arrived [4] in the 1950s, spurred by the introduction of mainframe computers, digitization, and eventually the internet [27], the industrial designer also being revolutionized in this. The second phase of ergonomics spurred by this revolution was driven by the emergence of usability [28], which in turn changed the role and focus of the industrial designer. The mix of electronics and objects, for example, in radios, TVs, and video recorders led to more complex products, and so to the term usability becoming an important element in design. The development of the computer mouse is a good example. The first prototypes of a point and click device were constructed in 1963. It, however, took many years before the mouse we know today was developed, the first generations being far from intuitive to use. Each mouse was supplied with a thick user manual [29]. Usability was therefore the precursor of interaction design [30].

Jumping to the 1960s and 1970s brings us to the introduction of a new technology that had a major impact on industrial design and the designer. This technology was the Computer Numerical Control (CNC) machine, which allowed shapes to be milled, drilled, and lathed automatically from a coded control system. Digital drawings prepared on computers combined with CNC gave a much higher degree of accuracy and repeatability. The molds for car exteriors were, prior to CNC, handmade, the shapes often being voluptuous, intricate and including huge curves [31]. If you, however, looked very carefully at these handmade molds you would see that the left side was not the same as the right. Making the molds was also quite time consuming.

The introduction of CNC meant that the designer could now transfer drawings directly to production, and could be confident that the molds would be accurate

Fig. 3 A Beetle and a Cortina (Pictures from family albums)

and perfectly symmetric [32]. This transition in design and production techniques was something that consumers could see in the products of the time. CNC had a number of limitations in its infancy, double curved surfaces being for example difficult to program [33, 34], the designer therefore suddenly being limited in scope. The introduction of these limitations can be clearly seen in the ending of the iconic Volkswagen Beetle story [35], a truly voluptuous, curvy car and the introduction of the next generation, the Golf. There were almost no double curved surfaces on the Golf, the expression of the Golf being, of course, not just due to production limitations but also trends and other user needs. Other cars introduced at this time also showed a similar trend, double curves being replaced by hard lines and flat surfaces. Take a look at the Ford Cortina shown in Fig. 3. The pictures of the Beetle and the Cortina, which are from a family album, were taken in the same year. The designs of the two cars are, however, very different.

The technology of today does not, however, limit shape in the same way, double curved shapes being again found on cars. Some companies still today make a full-scale clay model [36]. However, the digital models and not molds taken from the clay model as previously, are used in production. Computer models provide many opportunities, but also have limitations, which industrial designers still today need to understand. Computer modeling programs, despite continuous improvement, still have shortcomings, in particular with the meetings between different shapes. This type of extremely complicated mathematics can be performed successfully by the human hand and eye, but not necessarily repeatably. The designer therefore has to be able to both give the product the desired shape, but also understand what can be digitized, where the limitations to digitization lie.

Many other factors other than technology have led to changes in the industrial designer profession [37]. These include trends, new laws, a more globalized world, and social media.

The third phase of ergonomics is based on the concept of designing for pleasure, emerged in this period [38]. Design transitioned in this phase to focusing on satisfying needs. Not just basic needs, but also needs at a metalevel. Products were now, for example, required to also please and create enthusiasm, this optimization to a more

Fig. 4 Door opener with new and old button (Photo Steinar Killi)

expansive user experience lifting the importance of interaction design. This can be exemplified by looking at how the term universal design has changed design and architecture.

North Carolina State University in 1996 defined the term universal design [39], which changed design focus from designing for the average user to designing for those who did not fit under the label "average" [40]. Designing also for those who are for example shorter, taller, or who have disabilities, led to a change in the design process of both industrial designers and architects. The philosophy behind this approach is that taking a wider spectrum of users into consideration would benefit everyone, those for example with disabilities but also those who could be labeled average [41]. The introduction of this design approach can be clearly seen around us today, one example being the backdoor of a university built in 1999–2001, as shown in the picture below (Fig. 4).

Fig. 5 Double railing, to cater for the height differences of people (Photo Steinar Killi)

The door was first built without taking into consideration universal design, as shown by the white switch for opening the door in the picture. The building was, however, later refurbished to meet universal design standards, and a new button for opening the door was installed. This button is much longer than the original one, so a wider group of people can reach it. The old switch was, interestingly, not removed, as this would have been expensive and would have required the whole doorframe to be changed. The door therefore has two working door openers, these aptly providing a visual representation of the adaption of designs to new insights and laws. Figure 5 shows a double railing that caters to people who are taller or shorter than average.

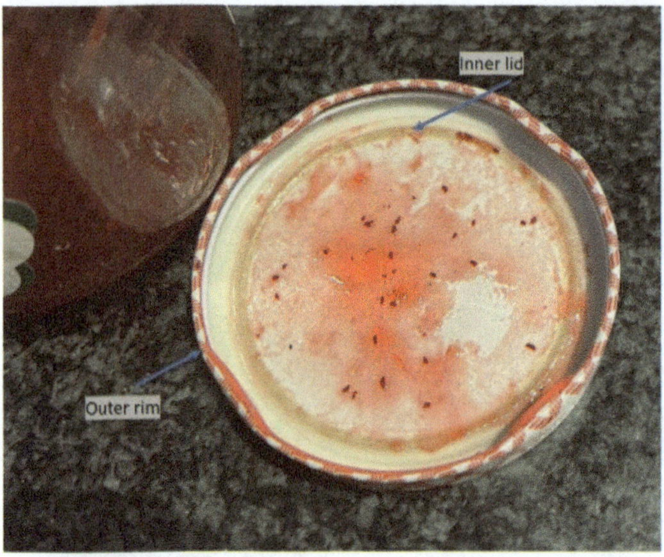

Fig. 6 A smart lid for a jar of jam, two parts that rotate and make it easy to open

Figure 6 shows a quite genius lid for jars of jam, it consists of two parts, an inner lid and an outer rim, the rim rotates around the lid, making it easy to open, for all.

Universal or inclusive design therefore aims to provide solutions that not only help those with disabilities, but provide better products for everyone [42]. Other examples include electric toothbrushes, low entrances to trams and buses and wider doors to shops and bathrooms.

The industrial designer has, throughout these industrial revolution periods and all the changes they represent, been designing products for mass production. Is the next revolution in design, however, a move toward customized mass production?

Digitalization is the prerequisite of the fourth industrial revolution [2], which builds on the third. The complexity of the artificial intelligence, augmented reality, robotics and 3D printing [43] involved in this digitization means that this revolution will take longer. It will, however, change how people create and live their lives, and how we create value. The industrial designer will also be changed by this revolution, perhaps more than ever before. The revolution has just begun.

How additive manufacturing can challenge and transform the term industrial design.

2 Additive Manufacturing

Additive manufacturing was introduced at an auto show in Detroit in 1988 [44]. The technology at that time was labeled stereolithography (see Fig. 7) and involved the use of a laser to harden resin built up layer by layer based on a digitized model constructed in a Computer Aided Design (CAD) program. The technology was initially developed to build models quickly, and with high precision, and therefore led to the term Rapid Prototyping, other AM technologies coming to market in the following years. SLS (Selective Laser Sintering), which used a pulverized nylon powder, also used a laser to fuse/sinter the plastic powder layer by layer. Support structures were not required, as the powder around the sintered part provided perfect support, the finished parts being excavated from the powder cake as shown in the picture (see Fig. 7). Another method used plastic coiled cord, the plastic being fused in layers like toothpaste, as shown in the picture (Fig. 7). This technique is called Fused Deposition Modelling (FDM) and is the most widely used, this being the technology found in private homes and schools. Other technologies have also been developed, and I will return to the interesting ones later. Rapid prototyping found its place, in the 90s and beginning of the 2000s, in companies that develop products. All car companies had, for example, rapid prototyping workshops with a number of machines, the ability to quickly produce precise models being a tremendous help to designers. Decisions could also be made more quickly [45], making it easier to involve stakeholders in the design process. The technology was primarily used to build prototypes. The first time this technology was used to produce a part for actual use is, however, difficult to determine. Additive manufacturing was probably used for the first time to produce a spare part that was difficult to source. The spare part could have been quite simple, as in the pictures below of knobs and details for an old drawer (Fig. 8).

The technology was given many different names over the years, including rapid prototyping, the term evolving into rapid manufacturing when the idea of using the technology to produce parts emerged, finally becoming AM. We should also not

Fig. 7 From left SLA (Stereo Lithography), SLS (Selective Laser Sintering), and FDM (Fused Deposition Modelling) (Photo Steinar Killi)

Fig. 8 Example of spare parts for an old drawer (Photo John Bugge)

forget 3D printing, which is probably the most widely known term and one that visualizes the technology best [46].

The use of additive manufacturing technology started to mature at the beginning of the 2000s, the ability to produce very complex parts, even unique parts, opening up new opportunities. Artisans or craftspeople were some of the first to use this technology to produce commercial end user products, a famous example being the Trabecula bench designed by Janne Kyttänen. Not many copies of this bench were made, and the price was high. It, however, served as an inspiration and example of the opportunities AM could provide. Similarities can be drawn between this and the Wassily chair designed by Marcel Breuer. Not many of the chairs were also sold. It, however, paved the way for using free form technology creatively, and showcased the possibilities of the technology. The picture below shows a very early example of the artistry of the artist and graphical designer Marius Watz. A digital "swirl" has been given physical properties and been produced using an indirect metal infusion method from a sintered wax model of steel particles infiltrated with bronze. Five examples of this artefact were made, and were awarded for the best engineering achievement in Norway from 2007 to 2011.

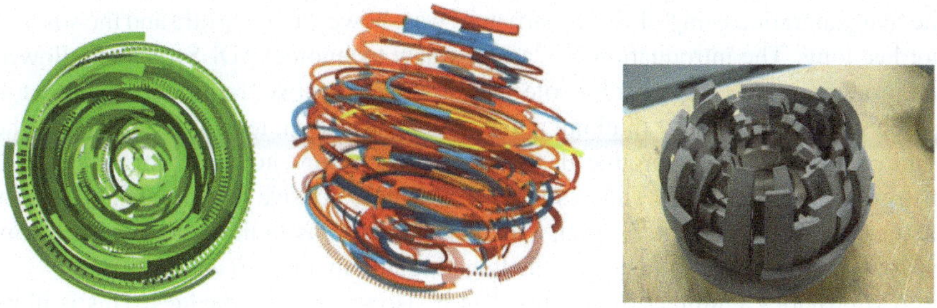

Fig. 9 Left and middle, the digital swirl that's been made into a solid part, last picture, Design Marius Watz (Photo Marius Watz and Steinar Killi)

The above (Fig. 9) shows the movement of artists, via technology, into the design role. Can we, however, also see the engineer? The answer to this is a very clear yes. Let's take a look at probably the most famous industrial AM project, in the hearing aid industry. A number of companies, including Siemens, around 2004 developed a method for customizing hearing aids [47]. Only a limited number of hearing aids were available at this time, users having to pick the one that fit best. This was not a good solution, as ears are complex and differences between people can be great. Picking a hearing aid from the small number available is also quite a long way from the universal design credo. AM has, however, provided a perfect solution to hearing aid design and production. Hearing aid components are small, complex, and expensive, therefore becoming the benchmark for other entrepreneurs considering AM as a manufacturing method. The process has been described in a number of texts [48] and is briefly as follows. A mold of the patient's ear is taken by a doctor in silicon, and is sent to the production facility where the mold is scanned. The digitized model is then hollowed to make space for the electronics, and then printed. The electronics are installed and the complete hearing aid is then shipped back to the doctor and patient as a unique, customized product. AM as a manufacturing method, and its benefits, have been showcased by both artists and engineers. How would, however, industrial designers use AM?

Very few consumer products were produced using AM for quite some time, these products also being at the experimental and discursive end of the scale. There are many reasons for this, the most widely cited being the shortcomings of the technology. AM was still too slow, too expensive, not accurate, or repeatable. The materials required were also in short supply, and there were quality issues, both mechanical and aesthetic [49, 50]. These issues are slowly beginning to be addressed, new AM technologies seeking to improve all areas of the technology. We can turn to Adidas, a super brand that produces millions of shoes each year, for an AM example. Adidas had been experimenting with AM as a production method for a special line of sneakers, the Futurecraft [51], for around 10 years. The idea was to design a sole with a structure that was customized to the user, stiffer in some areas and softer in others, the first generation using SLS. The structure was, however, too brittle and the sole did not last long. The introduction of Carbon's new technology (DLS), which allowed the production of a high-quality sole within an hour or two, represents the point at which Adidas could start thinking about mass production, tens of thousands of the Futurecraft 3 shoe being produced, see Fig. 10. They were, however, not customized, just produced using AM. This resulted in a quite expensive product, but with an x factor of being "3D printed" [52]. This neatly brings us to the second reason why custom products have not become the huge hit anticipated two decades ago.

AM provides a number of benefits. But it also presents challenges. This is illustrated in the figure below, Fig. 11, which shows an experiment we conducted some years ago.

The manufacturing costs of a product are an important component of total costs, and should therefore be taken into consideration when developing a new product. Saving a few cents on materials or production time can lead to huge savings and improved earnings, but only where a very large number of components are produced.

Fig. 10 Adidas Futurecraft. The sole is produced using DLS (Digital Light Synthesis) from Carbon (Photo Steinar Killi)

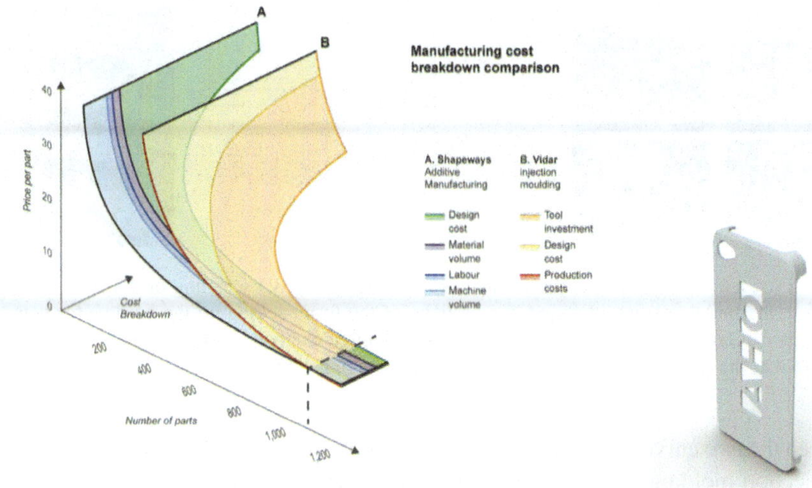

Fig. 11 Comparison of cost, development, and production of a phone cover using AM and injection molding (Drawings William Kempton)

Tooling costs can also be written off across the many components produced. A tooling cost of $100,000 for the production of 100,000 parts gives a tooling price per product of $1. Materials, machine, and development costs are, of course, also in addition to tooling costs. Development costs are a significant contributor to total costs, but can be spread across the total number of parts produced. If the development cost of a component is $100,000, then this adds another dollar to the total costs per product where 100,000 are produced. There are no tooling costs in additive manufacturing, so saving this cost of $100,000. Parts can therefore be produced at a lower cost, pure production costs per unit being independent of volume. Development costs have not, however, evaporated.

The product must still be designed and the costs associated with this must be spread across all the parts produced. If development costs are still $100,000 and just 100 parts are produced, then each part will have a start cost of $1000. As we can see, AM production costs of $10–50 or even $100 therefore are just a small proportion of total costs, which includes development. This situation can be represented by two curves, as shown in the example above, Fig. 12. The two curves show the total cost (all costs included) per part as a function of the total number produced, the first curve being for AM, and the second for injection molding The cost for producing just one phone cover using AM is much lower than where injection molding is used. It will, however, be far from cheap, the phone cover being very expensive and also quite simple. The figure therefore shows that injection molding should be used if we expect to produce more than around 1,100 copies. These curves would be different if the cover was more complex and the tools were more expensive, the intersection

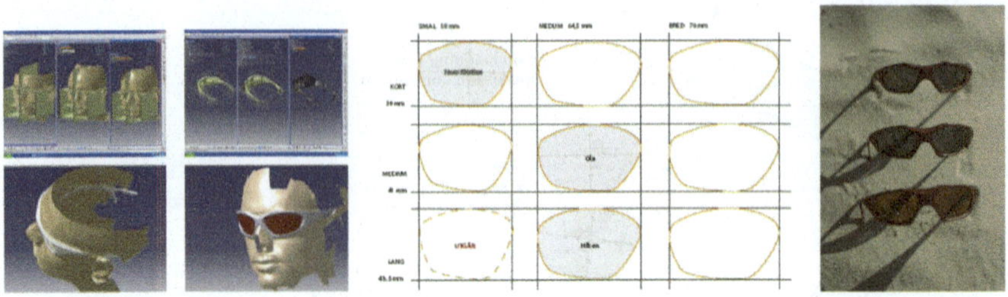

Fig. 12 First left shows how measurements were derived from a 3D scanning of the customer's head/face, the middle picture showing the elasticity of the form, setting up the borders for how the design can be stretched. The last (far right) shows three examples, all different in size, but recognized as the same brand (Design and photo Kathinka Bene Hystad)

between the two curves moving to the right and more covers needing to be produced for injection molding. This, however, is the point at which the industrial designer comes in, irrespective of whether designing for mass production, designing for mass customization or designing unique products for limited number runs for a high-end market. The industrial designer moves toward craftsmanship for the high-end product, becoming more of an artisan, and requiring a new design approach for this new paradigm.

Mass customization did not of course start with AM [53]. There have always been options when buying products. The famous model T-Ford introduced by Henry Ford was available in different colors when launched. It was, however, not long before it was discovered that black paint dried fastest, all cars then only being available in black to increase production speed. This change led to the famous quote by Henry Ford of: "people can have the car in any color they want, as long as it is black!" Today you can customize your car in many, many ways, the total number of combinations of, for example, color, interior, and engine for each model often extending into the millions [54]. Today's assembly lines cater for all these variations, a visit to a car manufacturing plant showing cars of different colors and with different interiors all moving along the same assembly line, the different parts being delivered to the right car through the use of an extremely complex logistic system. The designers have facilitated the different options. They are, however, out of the loop when manufacturing begins.

Hearing aids and Adidas are examples of mass customization. The true potential of AM as a tool for industrialized customization is, however, often not realized. Adidas has chosen to continue to produce standard sizes, even though the production technology can provide individualized shoes. A large amount of data would, however, need to be collected and processed before AM could be used to generate a customized sneaker.

3 The Present

3.1 Custom Sunglasses

Another example is the sunglasses developed by a master's student at Oslo School of Architecture and Design in 2005 for mass customization. The pictures below, Fig. 12, show the process.

The process begins with a start design that can be twisted and customized based on data derived from a face scan of the customer. The sunglasses frame dimensions could be stretched and tweaked to the width of the nose, and the distance between a number of major points on the face. The customer could finally make small adjustments to the frame before being produced using AM. This example not only provides a functional product of perfect fit, but also one that aesthetically fits the personal taste of the customer, within a predefined limit. The designer made the bridge of the sunglasses, the Y, the recognizable branding element, an element which could also be tweaked to absorb changes, so catering for an elasticity of form [55]. The designer developed a template which allowed users to adjust the glasses for fit, but also to their personal likings, the overall expression of a distinct brand through this also being secured. A number of AM glasses have been subsequently introduced by leading designers, but not as a customizable product. This example illustrates the emergence of interaction design and its morphing into service design, a number of touch points allowing the user to influence the solution, even at different levels. The design process then becomes part of the service, which is strongly connected to the product. We also see this in kitchen solutions. A kitchen can relatively easily be planned, drawn and so customized to a specific room and to the buyers' specific needs and desires.

The insights gained from this student's work and from the work of others have led to the development of an AM design approach. I will, in the following, give an introduction to this method, which has been covered in more depth in the book "Additive Manufacturing, design methods and Processes" [46].

3.2 The AICE Approach

The AICE approach attempts to address the complete design process, not just the final preparation of a design for AM (a misconception in society about how AM should be best used). A number of companies have, over the years, approached me with concepts for new products, most being almost ready for production. Their question usually is "Can you produce this more cheaply than injection molding?" My answer usually is no. The product has been designed for a conventional production method, optimized (for example) for injection molding and has a price per part of less than $1. AM should, in other words, have been made the production method from the very beginning, and before major decisions were made. This would allow the designer to base the design on the possibilities and challenges/problems of AM. Figures 13 and 14 show how this change influences the product development process.

Conventional production processes typically lead to a linear development path, starting with design, moving through engineering before finally arriving at production. There are overlaps and the designer is very often a part of the whole process, often coming to the factory to watch the first products being produced. This can sometimes lead to (hopefully) small changes, made either directly at the production site, and sometimes offsite by the engineers. The plastic used may not be the perfect choice or the screws need to be adjusted. It is, however, extremely rare that the process must be returned all the way back to the designer and the design process started again. This process can be described as a trickle-down process (see Fig. 13). The introduction of a production method that does not involve high tooling costs suddenly, however, provides an opportunity to challenge this process. A product can move faster through the process, so allowing it to be produced for testing in the market and for feedback to go all the way back to the design, for incorporation into the final version. This very much resembles computer software development, the product quickly getting to market, the design then being updated. This could be labeled a trickle across process (see Fig. 14), industrial designers building on methodologies first developed by interaction designers.

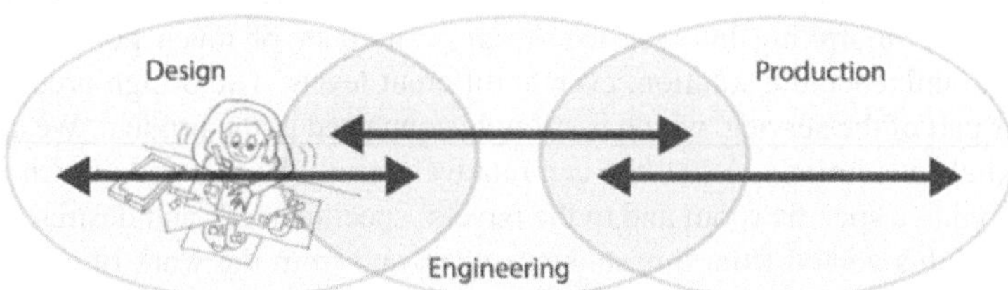

Fig. 13 Trickle down, the classic linear (waterfall) process. Interactions between the different steps toward production. The designer, engineers, and manufacturers are, however, assigned quite specific roles (Drawing Inger Steines, Stein Rokseth, Steinar Killi)

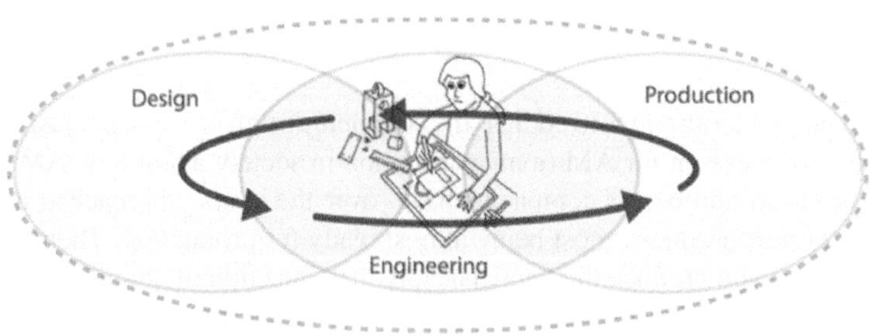

Fig. 14 Trickle across. The designer moves to the center of the product development model, feedback continuously trickling across the full width of the model (Drawing Steinar Killi, Stein Rokseth, and Inger Steines)

3.3 The AICE Method

The AICE method is thoroughly presented in [46]. An outline of the method and some examples are, however, given below. AICE is an acronym for Adapt, Integrate, Compensate, and Elongate, which are the pillars of the method. These pillars influence each other, but in succession and in the direction of the acronym. I will give a short introduction to each below, starting with the first, Adapt.

Adapt. The first pillar of the method addresses the analytical and creative stage, an example being the adaption of existing design methods to the new opportunities provided by AM. AM gives a very high level of freedom of form. Methods such as forced relations [46] can be used or value through extra functionality can be added to a product at almost no extra cost. The pictures below (Fig. 15) are examples, taken from a project in which design students were required to develop phone covers for AM production, which were sold for a short period on the website http://fabrikkaho.no/.

Rasmus Agerup developed a quite comprehensive phone cover with a lot of extra functionality, including a cardholder, an adjustable support for watching movies and personalization. All the features were developed early in the process, as shown by some of the pictures from the process.

The next pillar, Integrate, relates to designing for additive manufacturing. This field has been covered widely, especially in white papers from AM machine developers. This stage focuses on features that can change with production method [56]. For example, if method is changed, will any support structures need to be removed? Is there a minimum wall thickness? The phone cover sample was developed for SLS, which allowed rotational hinges and a thin pocket to be designed and produced. However, the use of the SLS method required how the loose powder would be removed to be considered, the design therefore avoiding pockets where powder would be hard to remove.

Compensate. Different AM processes have different limitations and challenges, which must be addressed [57]. Accuracy will vary with technology, which however

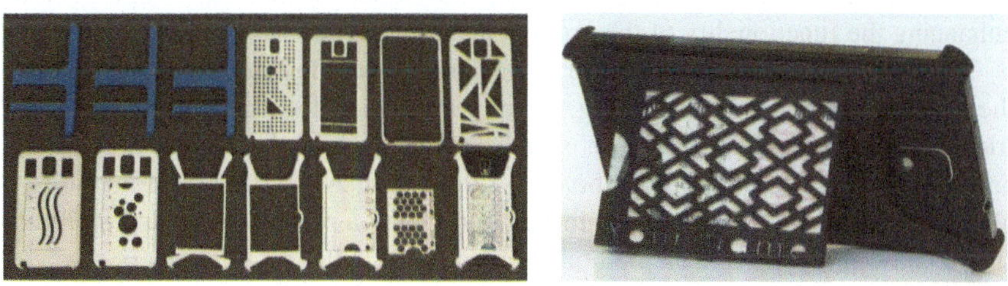

Fig. 15 Left, images from the design process. Right, the actual result (Design and photo Rasmus Agerup)

is, despite continuous improvement, still an issue [58]. So is cleaning and post-processing. Compensate involves the integration of these limitations and challenges into the design, this maybe also acting as a feature. Perfect fit of phone covers is an AM challenge, covers easily being a little too big or a little too small. Some flexibility is also needed to achieve a snap functionality, and it should also be possible to remove the cover. We can see, looking at the cover in Fig. 15, patterns of holes in the surface that give the cover the elasticity and snap fit to the phone required. These patterns can, however, also be used to add value, through using them to create personalized symbols, letters, or even pixelated pictures. This requires an industrial designer to hold core competencies such as a wide understanding of materials, their aesthetical possibilities, and mechanical properties. It also requires an understanding of materials beyond the material, including light, RFID fields, and sound.

This production method does, however, involve a paradox. The method favors low volumes and more precise and unique/customized products. The low volumes, however, mean that development costs will, as described earlier, be huge per part.

Elongation, the last pillar is the solution to this, products being designed for an elongated life. There are many ways of achieving this. We described hearing aids earlier, the production system used allowing the design to be tweaked, so opening up for customization at an industrial scale. The hearing aids are, however, not really designed, the design being just a mold of the patient's ear. The process and fitting of electronics is, of course, designed. The actual shape of the hearing aid is, however, given by the ear and not by a designer. The sunglasses are a more relevant example, design involving two key factors. The first is the development of a design that can be tweaked within set limits, each measurement having minimum and maximum values. The final design is therefore generated from a scan, measurements being made in a CAD program. The visual appearance of the sunglasses, unlike the hearing aids, is a vital part of the sunglasses' value. The designer furthermore introduced a branding element, the "Y" bridge between the glasses, to allow new branded editions of the sunglasses. This also serves as a recognizable element that links the different sunglasses and opens up for new versions. This mix of permanent elements and adjustable components and areas, framed by the designer, allows a potential user to cocreate the design. The user therefore moves deeper into the design process, enhancing the functionality of the product and as an experience.

The design process of the phone cover and the sunglasses differ slightly from the classical process depicted in Fig. 13, migrating from the classical process to the trickle across model. The designer is much more involved throughout the process, and the path back to the drawing board is quite short. AICE, like most methods, is also about disrupting an existing mindset, and giving a process focus and direction. If you know and understand a method, then you can challenge, improve, or change it at your discretion, this becoming clearer in iterations, especially where real changes are made, not just superficialities such as change of color.

4 The Future

AM is not a technology of the past. Nor are processes such as injection molding, casting, or other manufacturing methods. They will all continue to be enhanced and optimized for many years to come, the majority of products being manufactured using classical production methods. New technologies will also emerge. Let's take a look at some that we can envisage being implemented, tested, and developed.

This chapter began with a description of the different industrial revolutions, one of the instigators of the fourth revolution being claimed to be AM [52]. There are, however, other instigators including nanotechnology, artificial intelligence, human–machine interface, advanced digital production, and complementary and other technologies [59]. An example from the maritime industry showcases this well, it being from a huge research project OpenVr, which is closely connected with industrial partners such as Vard. The project seeks to use technology, develop technology and converge/compliment technology closely with the users, mainly professionals working on large ships and in a very complex working situation [60]. The tools used include virtual reality and augmented reality, which give designers the opportunity to work closely within the environment they are designing for, and to test in real life, on a ship bridge and in a simulator, the interaction between the designers and the users being a vital element in the development of innovative solutions.

We see that a mix of design techniques and materials (both physical and non-physical) enables a process in which advanced technologies can be paired with direct user involvement. This will speed up the design process, lower risk, and therefore costs. The situation can in itself, as shown in Fig. 16, be extremely complex, but can be carried out in a plain room using VR (Fig. 16). Bringing users into this situation therefore represents an additional challenge. An important tool for industrial designers can, in the near future, be more experiential methods such as augmented reality. This could bring the users even closer to the development process, in real time.

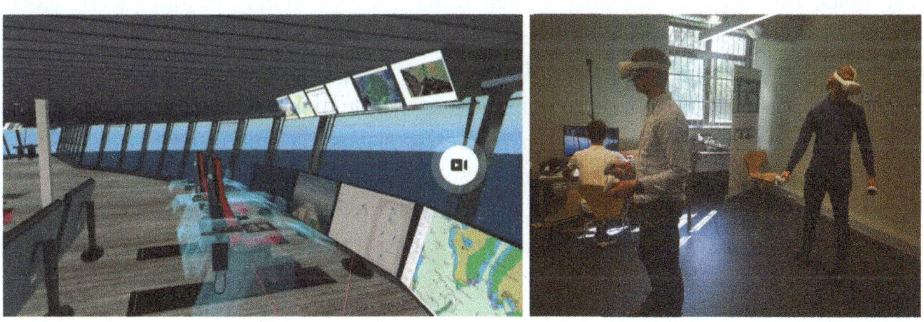

Fig. 16 Designers move inside a virtual environment, a bridge on a huge ship. They can test out different solutions, and interact directly with potential users, the ship's crew (Photos Etienne Gernezs and Jon Erling Fauske)

Technology clearly influences the designer, the process, and the result. There is, however, a growing understanding of the role of the user in this development path. At the start of this chapter we can see that the designer had a user in mind. Henry Ford, however, was focused on producing as many cars as possible, removing user/customer options in his endeavor to achieve this goal. Tupperware was, alternatively, developed and flourished through direct contact with the customer, feedback from Tupperware parties giving designers much greater insight than selling off the shelf. AM is, however, on the threshold of taking this further, customizing on a personal scale now seeming to be within reach. What we can anticipate from the near future is new technology not just making the world more complex, but also bringing the user closer to the design process, and to the technology.

5 Summing Up

The intention behind this chapter was to provide both an insight into how the role of the industrial designer has changed since emerging almost exactly 100 years ago. The role of the industrial designer emerged from the introduction of new technologies and from changes in how we look at users of the products designed. The emergence of a disruptive technology like additive manufacturing can also be a gamechanger for the industrial designer. These changes do not just involve the adoption of a new technology, but also the interpretation of how this can change the relation to users. Looking at universal design and how we can develop an environmentally, but also economically and societally sustainable future, it becomes obvious that we need to think and act differently. This is also the reason behind introducing the idea of different revolutions. If AM is going to have the impact that would justify the term revolution, then greater value has to be created by producing smarter and viewing the users as both people and the planet [61]. This could be achieved by products truly being a part of a circular system, being desirable and useful products, made on demand, closer to the customer, and of materials that can be easily recycled/reused when the lifetime of the product comes to an end. This can also be achieved by engaging and connecting different groups in different product development processes. This is depicted in Fig. 17 as possible distributions between different stake holders, from home designer/prosumer, to artisans, to big international companies. The top panel shows a possible distribution of production between stakeholders, the lower panel showing how design could be distributed between stakeholders. There are, as we can see, many possibilities for changing the entire product development process.

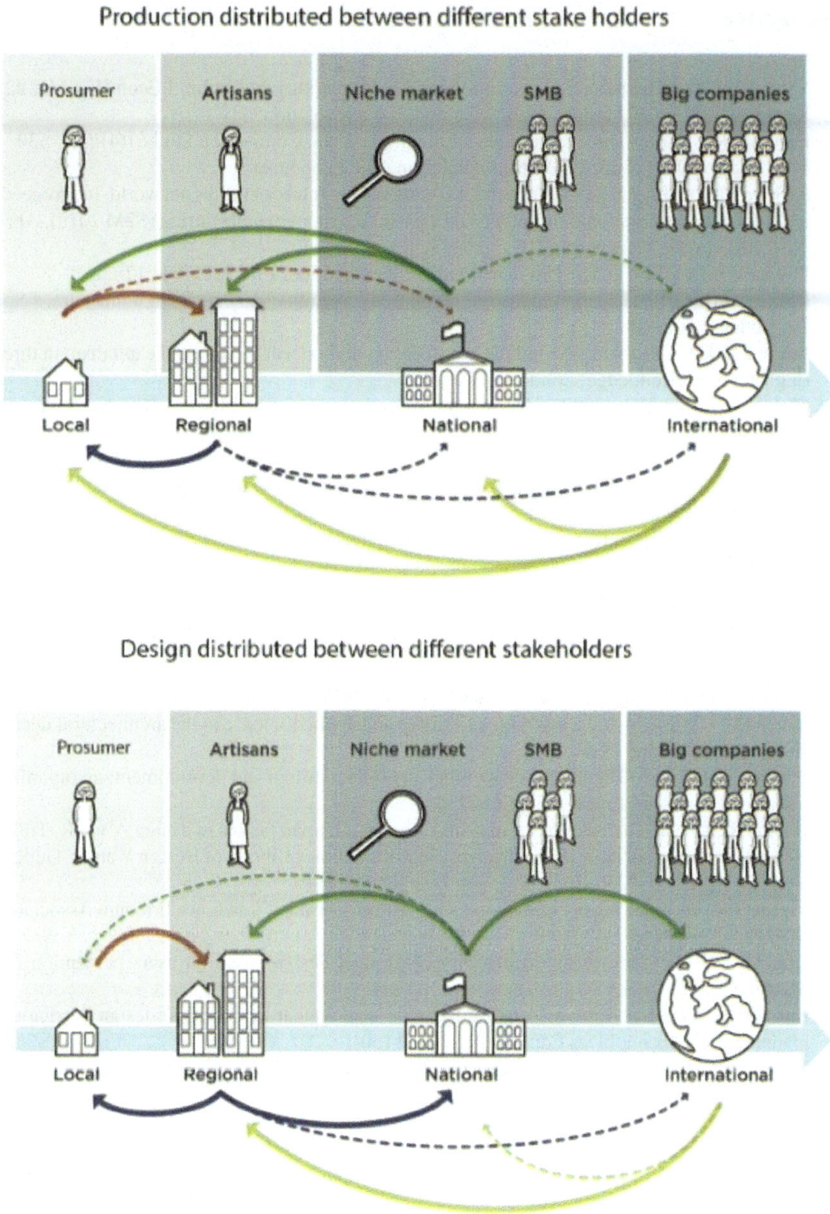

Fig. 17 Top panel. Different possibilities for distributing production between different stake holders. Lower panel. Distributing design between different stakeholders (Drawings Inger Steines and Steinar Killi)

References

1. Vries JD (2009) The industrial revolution and the industrious revolution. J Econ Hist 54(2):21
2. Klingenberg C (2017) Industry 4.0: what makes it a revolution?
3. Easterlin RA (2019) Three revolutions of the modern era. Comp Econ Stud 61(4):521–530
4. Markillie P (2012) A third industrial revolution. The Economist
5. da Silva Bartolo PJ et al (2020) Industry 4.0–shaping the future of the digital world. In: Proceedings of the 2nd international conference on sustainable smart manufacturing (S2M 2019), April 2019.CRC Press, Manchester, UK, pp 9–11
6. Philbeck T, Davis N (2018) The fourth industrial revolution. J Int Aff 72(1):17–22
7. Mohajan HK (2019) The first industrial revolution: creation of a new global human era. J Soc Sci Humanit 5(4):10
8. Foster J (2003) Class struggle and the industrial revolution: early industrial capitalism in three English towns. Routledge, London
9. Mokyr J, Strotz RH (2000) The second industrial revolution, 1870–1914. Northwestern University
10. Lee J, Lee K (2021) Is the fourth industrial revolution a continuation of the third industrial revolution or something new under the sun? Analyzing technological regimes using US patent data. Ind Corp Chang 30(1):22
11. Lees-Maffei G, Houze R (eds) (2010) The design history reader. Berg, New York
12. Hounshell D (1984) From the American system to mass production, 1800–1932: the development of manufacturing technology in the United States. Johns Hopkins University Press, Baltimore
13. Frayling C (1993) Research in art and design. Royal College of Art Research Papers (1:1), pp 1–5
14. Cross N (2000) Engineering design methods, 3rd edn. Wiley, Chichester
15. Schön DA (1998) Toward a marriage of artistry and applied science in the architectural design studio. J Archit Educ 41:6
16. Slappendel C (1994) Ergonomics capability in product design and development: an organizational analysis. Appl Ergon 25(5):266–274
17. Huston RL (1994) The measure of man and woman—human factors in design Alvin R. Tilley, Henry Dreyfuss associates 1993, 96 pages, $60.00. Whitney Library of Design Watson-Guptill, New York ISBN 0–8230–3031–8. Ergon Des 2(2):37–39
18. Meister D (1999) The history of human factors and ergonomics. Lawrence Erlbaum Associates, Mahwah, New Jersey
19. Rong W (2012) Study of the availability of product design based on users perception [J]. Machinery
20. Liu S-F et al (2013) Applying quality function deployment in industrial design curriculum planning. Int J Technol Des Educ 23(4):1147–1160
21. Waterson P, Sell R (2006) Recurrent themes and developments in the history of the ergonomics society. Ergon 49(8):743–799
22. Batchelor R (1994) Henry Ford, mass production, modernism and design. Manchester University Press, Manchester
23. Crowther S (1922) My life and work: an autobiography of Henry Ford. William Heinemann Ltd
24. Weber NF (2009) The Bauhaus group: six masters of modernism. Alfred A Knopf, London, p 544
25. Clarke AJ (1999) Tupperware: the promise of plastics in 1950s America. Smithsonian Books, London
26. Clarke AJ (2014) Tupperware: the promise of plastic in 1950s America. Smithsonian Institution, London
27. Greenwood J (1997) The third industrial revolution: technology, productivity, and income inequality. American Enterprise Institute, Washington, DC

28. Jordan PW (2020) An introduction to usability. CRC Press, London
29. Atkinson P (2007) The best laid plans of mice and men: the computer mouse in the history of computing. Des Issues 23(3):16
30. Hornbaek K, Stage J (2006) The interplay between usability evaluation and user interaction design. Intern J Hum-Comput Interact 21(2):117–123
31. Tumminelli P (2004) Car design. Te Neues Verlag GmbH
32. Dowlen C, Shackleton J (2003) Design history of the car: an empirical overview of the development of layout and form. In: DS 31: Proceedings of ICED 03, the 14th international conference on engineering design, Stockholm
33. Raun C, Kirkegaard P (2015) Adaptive mould-A cost-effective mould system linking design and manufacturing of double-curved GFRC panels. In: 17th international congress of GRCA-GRC, Dubai
34. Simondetti A (2002) Computer-generated physical modelling in the early stages of the design process. Autom Constr 11(3):303–311
35. Barber C (2003) Birth of the Beetle: the development of the Volkswagen by Ferdinand Porsche. Haynes Publishing, Sparkford
36. Tovey M, Owen J (2000) Sketching and direct CAD modelling in automotive design. Des Stud 21(6):569–588
37. Micheli P et al (2012) Perceptions of industrial design: the "means" and the "ends." J Prod Innov Manag 29(5):687–704
38. Green WS, Jordan PW (2002) Pleasure with products: beyond usability. CRC press, London
39. Story MF (2001) Principles of universal design: universal design handbook. McGraw-Hill, New York, NY
40. Carr K et al (2013) Universal design: a step toward successful aging. J Aging Res 2013:324624–324624
41. Bergman E et al (1996) Universal design: everyone has special needs. In: Conference companion on human factors in computing systems
42. Mustaquim MM (2015) A study of universal design in everyday life of elderly adults. Procedia Comput Sci 67:57–66
43. Hopkinson N, Hague RMJ, Dickens PM (2005) Rapid manufacturing: an industrial revolution for the digital age. Wiley, Chicheste, West Sussex, p 285
44. Pham DT, Dimov SS (2001) Rapid manufacturing: the technologies and applications of rapid prototyping and rapid tooling. Springer Verlag, London
45. Capjon J (2004) Trial and error based innovation. In: Institute of industrial design. Oslo School of Architecture and Design, Oslo
46. Killi S (ed) (2017) Additive manufacturing: design, methods andf processes, 1st edn. Pan Stanford, Singapore
47. Reeves P, Tuck C, Hague R (2011) Additive manufacturing for mass customization. Mass customization. Springer, London, pp 275–289
48. Campbell RI, Jee H, Kim YS (2013) Adding product value through additive manufacturing. In: DS 75-4: Proceedings of the 19th international conference on engineering design (ICED13), Design for Harmonies, vol 4: product, service and systems design. Seoul, Korea, 19–22 August
49. Singh S, Ramakrishna S, Singh R (2017) Material issues in additive manufacturing: a review. J Manuf Process 25:185–200
50. Maidin SB, Campbell I, Pei E (2012) Development of a design feature database to support design for additive manufacturing. Assembly Automation
51. Yang H, Luo D (2019) A study of additive manufacturing technology's development and impact through the multi-level perspective framework and the case of Adidas
52. Galantucci LM et al (2019) Additive manufacturing: new trends in the 4 th industrial revolution. In: International conference on the industry 4.0 model for advanced manufacturing. Springer
53. Bardakci A, Whitelock J (2004) How "ready" are customers for mass customisation? An exploratory investigation. Eur J Mark
54. Alford D, Sackett P, Nelder G (2000) Mass customisation—an automotive perspective. Int J Prod Econ 65(1):99–110

55. Killi S (2010) Form follows algorithm: computer derived design for rapid manufacturing. In: Innovative developments in design and manufacturing: advanced research in virtual and rapid prototyping: proceedings of the 4th international conference on advanced research and rapid prototyping, Leiria, Portugal, 6–10 October 2009, pp 575–590
56. Page T (2011) Design for additive manufacturing. LAP Lambert Academic Publishing, Saarbrucken
57. Maleki E et al (2021) Surface post-treatments for metal additive manufacturing: progress, challenges, and opportunities. Addit Manuf 37:101619
58. Nazir A, Jeng J-Y (2020) A high-speed additive manufacturing approach for achieving high printing speed and accuracy. Proc Inst Mech Eng C J Mech Eng Sci 234(14):2741–2749
59. Schwab K, Davis N (2018) Shaping the future of the fourth industrial revolution. Currency
60. Aylward K et al (2021) Using operational scenarios in a virtual reality enhanced design process. Educ Sci 11(8):448
61. Silver N (2017) Finance, society and sustainability: how to make the financial system work for the economy, people and planet. Springer, Berlin, Germany

Using the Axiomatic Design in Engineering

Oana Dodun, Miguel Cavique, Laurenţiu Slătineanu, and Petru Duşa

Suh introduced two design axioms to the engineering community about four decades ago. Compared with other theories which presuppose an algorithmic or an iterative approach specific to a field of engineering, Axiomatic Design (AD) can be applied to solve a wide variety of problems. It is possible to apply AD to subjects in humanities and social sciences. Properly using those axioms leads to the best design solution for products, processes, or systems. The work briefly presents the AD approach in its three essential parts: axioms, structure, and process. Thus, this chapter presents a decision-making process using the first axiom. Then it illustrates the second axiom. The AD literature is analyzed, and the published papers were divided into theoretical developments or applications grouped by their field. The advantages and disadvantages of using AD compared with other design processes like QFD, TRIZ, and Taguchi are presented.

O. Dodun (✉) · L. Slătineanu · P. Duşa
Department of Machine Manufacturing Technology, Faculty of Machine Manufacturing and Industrial Management, "Gheorghe Asachi" Technical University of Iaşi, 700050 Iaşi, Romania
e-mail: oanad@tcm.tuiasi.ro

L. Slătineanu
e-mail: slati@tcm.tuiasi.ro

P. Duşa
e-mail: pdusa@tcm.tuiasi.ro

M. Cavique
Unidemi, Faculdade de Ciências E Tecnologia, 2829-516 Costa da Caparica, Portugal
e-mail: cavique.santos@escolanaval.pt

Department of Sciences and Technology, Naval Academy, Base Naval de Lisboa, 2810-001 Almada, Portugal

P. Kyratsis et al. (eds.), *Advances in Product Design Engineering*, Management and Industrial Engineering,
https://doi.org/10.1007/978-3-030-98124-2_2

1 Introduction

Engineering is defined as the practical application of pure science or the application of science to the optimum conversion of nature's resources to the uses of humankind. A scientific theory is applied to develop, design, and analyze solutions in all engineering branches. Over the years, there were proposed specific approaches for solving the wide variety of problems that respond to the questions "How to do things better?" and "How to do better things?" which engineers usually have to answer. Due to the practical aspect of the main goal, each solution has to satisfy conflicting requirements and safety requirements for a given cost. The significant functions of engineering are research, development, design, construction, test, modifying, installing, production, inspecting, maintaining, operating, and managing a wide variety of products or systems. Engineers manage to manufacture, supervise design and processes, recommend materials, conduct failure analysis, provide assistance services, etc.

Axiomatic Design (AD) theory was introduced to the engineering community about four decades ago. AD aims to avoid conflicting requirements. The idea is a consequence of students' project activities where they studied design processes and the quality of design solutions at the Massachusetts Institute of Technology conducted by NP Suh. This design theory was a real breakthrough because it can integrate all the main characteristics and ingredients of the design process. Due to its axiomatic approach, AD generalize and establish a common core for important design issues such as efficiency, robustness, simplicity, and probability of success. As Suh says, AD claims to establish a scientific basis for design [1]. Moreover, AD provides a scientific foundation for designers to compare the possible designs and choose the best option. The time and costs associated with the product development are minimized, and the customer needs are detailed.

For the first time, considerations regarding the possibilities of the existence of an axiomatic design methodology were to be published in May 1978 [2] discussing the axiomatic approach to manufacturing problems and manufacturing systems. In that first publication, several hypothetical axioms and some of their corollaries are presented with examples to illustrate the basic concepts. In 1990 Suh published his first book [1]. He explained the principles of design and emphasized that Axiomatic Design (AD) aims to make people more creative, prevent randomly looking for a solution, minimize the iterative process called "trial and error," and choose the best concept from the multitude of proposed concepts. In the early years, the cumulative number of publications was relatively low. After the first decade, the number of papers has significantly risen probably because of the second book published by Suh [3], in which he presented the AD advantages and applications.

The community with AD interest has been developed further, and in 2000 the first edition of the International Conference on Axiomatic Design (ICAD) was held in Cambridge (MA-USA). Since then, Cambridge and Worcester (MA-USA), Seoul and Daejeon (Korea), Lisbon (Portugal), Florence (Italy), Xi'an (China), Iasi (Romania), Reykjavik (Iceland), and Sydney (Australia) have hosted the ICAD conferences.

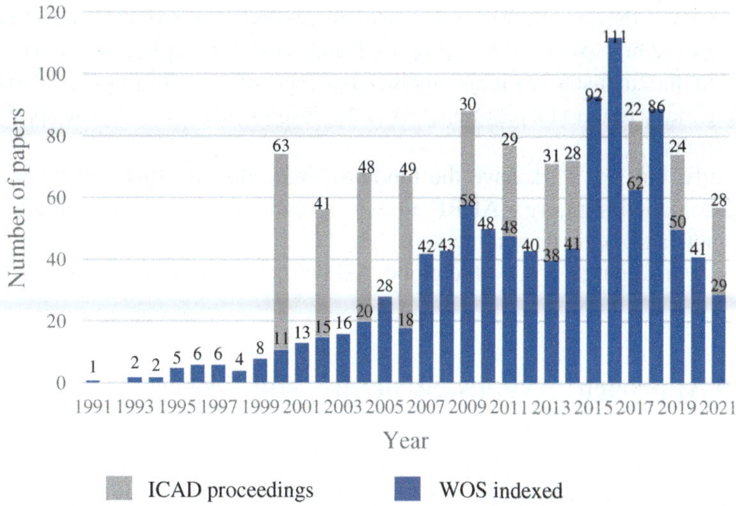

Fig. 1 Number of papers with AD topic, published over time in ICAD proceedings or indexed in Web of Science

Since 2013 the conferences have been held annually except for the pandemic restriction in 2020. Regarding the indexation of volume proceedings, the conference succeeded WOS indexation in 2015, 2016, and 2018. Figure 1 shows the papers published as ICAD conference papers in gray columns for the corresponding years. The blue columns represent the number of papers indexed in the Web of Science (WoS) database with the keyword Axiomatic Design.

To evaluate and classify these publications, few authors publish studies of literature review of AD publications [4–7]. In [4], Kulak et al. considered the papers published from 1990 till 2009. He classifies papers into four main groups, type of axiom used, application area, theoretical consideration on the method, and the type of evaluation. The study emphasized the discrimination concerning the used axioms in favor of the first axiom. The number of papers ranked the following application: Product Design, Decision Making, Software Design, System Design, Manufacturing System Design, and others.

In the form of a literature review, Rauch et al. [5] investigate the number and type of publications between 1996 and 2005 dealing with AD in Manufacturing. The data basis of this analysis is the works indexed in the Scopus. The study shows the preponderance of the first axiom vs. the second one and the application papers compared to the theoretical development.

A nomothetical empirical study of AD application in academic publication between 2013 and 2018 is presented in [7]. The conclusion is that the number of papers concerning system design grew significantly, but the number of papers concerning application for Software design has reduced.

With time, some private companies are involved in the AD community. Many projects have been conducted to promote, provide training, and implement the AD methodology in the design and development processes (www.axiomaticdesign.com). A software named Acclaro was proposed and used for implementing axiomatic designs in the industry with the main advantage of overcoming the barrier of managing many functional requirements of the design and facilitating the application of the axioms.

Recently, Dr. B.J. Park gave the funds to create the AD Research Foundation (www.axiomaticdesign.org). ADRF aims to spread knowledge, provide education and promote axiomatic design methods.

In some universities around the globe, AD is a course in the curricula.

2 AD Approach: Axioms, Structure, and Process

Axiomatic Design theory became a powerful tool for designing new products, systems or processes or analyzing and improving the existing design solutions. Due to its axiomatic approach AD apply to a wide variety of problems [1–6]. According to Merriam-Webster Dictionary, the meaning of the word "axiom" is "an established rule or principle or a self-evident truth," and its synonyms are "postulate" and "maxim." Like any other axioms in science, the validity of Suh's axioms, relies on the lack of observations to the contrary within their domains of applicability. Since 1978, when the first paper was published [2], nobody had come to prove that the axioms are not valid.

Brown presents [8] AD using three pieces: axioms, structure, and process. Each piece can be decomposed into two elements, as it is represented in Fig. 2. The design problems are formulated during the design process, and the design solution is developed, from abstract to detailed, across customer, functional, physical, and process domains structures. The zigzagging decomposition process happens between domains at one level, then proceeding to the next, more detailed level, constantly checking if the axioms are respected.

According to the AD theory, the entire design process of a new product or the improvement of an older design should be consistent with the following two axioms:

Independence Axiom (1.1 in Fig. 2)**: *"Maintain the independence of functional requirements."*** A product or system design is considered ideal if all functional requirements are independent of the others. The independence avoids any interaction among them, which can have unintended consequences. Without interaction, the functional design requirements are adjustable and controllable, and:

Information Axiom (1.2 in Fig. 2)**: *"Minimize the information content of the design."*** This Axiom helps choose among multiple possible solutions by favoring those with the greatest possibility of fulfilling the functions, maximizing the probability of success. Therefore the design satisfies the customer's needs.

AD theory proposes four design domains, **Customer Domain**, **Functional Domain**, **Physical Domain,** and **Process Domain**.

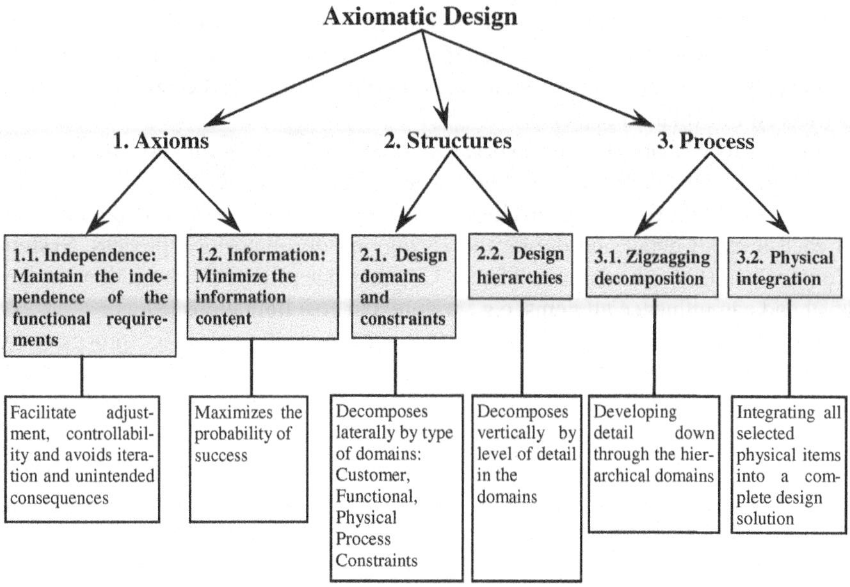

Fig. 2 Parts and elements of Axiomatic Design, after [8]

The Customer Domain consists of all the **Customer Needs (CNs)** or what the customers want or what is needed. A proper requirement gathering process or identification of the customer needs or other stakeholders' needs (manufacturing, transport, salespeople, etc.) is essential for the final design solution [9]. Significant CNs will be missed without recognizing all stakeholders, which probably leads to missed FRs and a less valuable design solution [9]. CNs should describe the fundamental needs, preferences, and constraints like cost, weight, volume, etc.

The Customer Domain maps into the Functional Domain. Its elements are the **Functional Requirements (FRs) that** describe the functions of the design solution. The FRs should state the design objective and should begin with verbs. The selection of good functional requirements (FRs) is essential for design solutions. According to Suh, "a design solution can be no better than its FRs" [1]. In [10] Thompson proposes a helpful classification of other FR-like entities, challenging to differentiate from real FRs. Such entities are the non-FR that describe the qualities or characteristics of the design solution, the *optimization criteria (OCs)* that indicate a maximizing or minimizing function, and the *selection criteria (SCs)* like cheapest, lightest, most robust, etc. The top-level FRs should be the minimum list of functions that satisfy all CNs which means the FRs should be *collectively exhaustive* concerning the CNs. FRs should be *mutually exclusive* concerning each other [8] at any level of decomposition. The top-level FR is FR0, decomposed at the next level into FR1, FR2, FR3, etc. The children of FR1 are FR1.1, FR1.2, FR1.3, etc.

In the Physical Domain the elements, the **Design Parameters (DPs)** describe how to implement the solution physically. The DPs are the physical items that fulfill the FRs. Ideally, the specific DP should influence only the FR that it is intended to satisfy. Moreover, the choice of the DP should meet the FR so that it ensures the highest probability to fulfill a requirement. In other words, the design to choose has the lowest information content.

In the Process Domain, the **Process Variables (PVs)** describe how the DPs are produced. Ideally, the PVs have a one-to-one relation with the DPs.

Developing **the design hierarchies** in all these domains by vertical decomposition at each level of detail and during the zigzagging decomposition process, **System Constraints (Cs)** are also considered. Cs introduces limits or restrictions at any level and can influence all items, e.g., weight and cost limitations [8].

Design hierarchies are established during the design process by mapping/**zigzagging decomposition** process between domains at one level, then proceeding to the next level (Fig. 3). Brown deals with the difficulties of developing good hierarchical decompositions [11]. Decompositions are used to solve problems when the solutions are not immediately obvious because the problems are too large or complex. In AD, the mapping runs the design across the domains. Vertical decompositions decompose the elements of a domain hierarchically and from abstract to detail. The children of a component must be *collectively exhaustive (CE)* with respect to the parents. The children must be *mutually exclusive (ME)* concerning each other at each level. Solving the horizontal and vertical puzzle raises a solution at a level of decomposition. In addition, considering the second axiom, the appropriate decomposition has the minimum information content, symbolized as *CEME-min*.

After the decomposition is complete, the selected items from the physical domain can be **physically integrated** into a complete solution.

The axiom of independence requires that the functions of the design (*functional requirements, FRs*) remain *independent*.

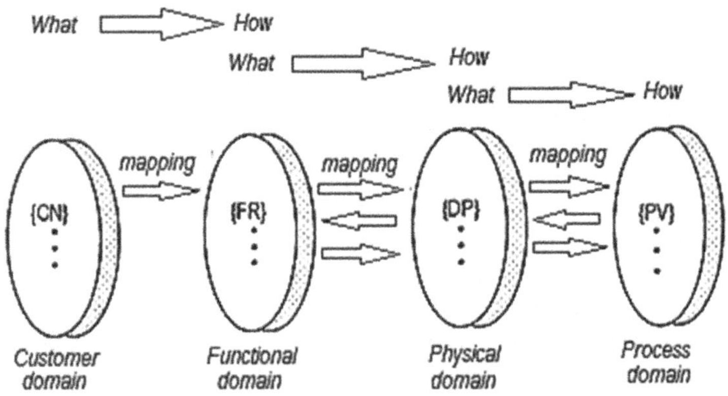

Fig. 3 Design activity approached as a process of "mapping" information, as a result of the transition from one field to another (Adapted from [3] and [12])

Subsequently, at a certain stage of application of the axiomatic design, it is necessary to get a physical issue to materialize each functional requirement. As aforementioned, the solution is the design parameter (DP).

The design matrix depicts the relations between functional requirements and design parameters.

Thus, *the general matrix relation* that connects the FRs functional requirements to the *DP*s design parameters has the form:

$$\{FR\} = [A]\{DP\} \tag{1}$$

[A] is the design matrix corresponding to a transfer function between the *FRs* functional requirements and the *DPs* design parameters.

In relation (1), A_{ij} is [1]:

$$A_{ij} = \frac{\partial FR_i}{\partial DP_j} \tag{2}$$

When there are *n* functional requirements, relationship (1) [1] has the form:

$$\begin{Bmatrix} FR_1 \\ \vdots \\ FR_n \end{Bmatrix} = \begin{bmatrix} A_{11} & \dots & A_{1n} \\ \vdots & \ddots & \vdots \\ A_{11} & \dots & A_{1n} \end{bmatrix} \begin{Bmatrix} DP_1 \\ \vdots \\ DP_n \end{Bmatrix} \tag{3}$$

Ideally, fulfilling a specific requirement by a DP_i design parameter does not affect any other functional requirements. It significantly increases the system's flexibility to modifications for improvements [13].

Matrix *A* allows to classify the type of designs:

(a) *A diagonal matrix* corresponding to the so-called *uncoupled design* (concepts). It has $A_{ij} = 0$, for all $i \neq j$:

$$[A] = \begin{bmatrix} A_{11} & 0 & . & 0 \\ 0 & A_{22} & . & 0 \\ . & . & . & 0 \\ 0 & 0 & . & A_{nn} \end{bmatrix} \tag{4}$$

When the matrix [*A*] is diagonal, each of the functional requirements *FRi* is met by a design parameter *DPi*. We will thus be dealing with *an uncoupled design matrix.*

(b) *A triangular matrix*, corresponding to the situation where there are non-zero elements either above the diagonal (*upper triangular matrix*) or below it (*lower triangular matrix*) [1, 14, 15]:

$$[A] = \begin{bmatrix} A_{11} & 0 & . & 0 \\ A_{21} & A_{22} & . & 0 \\ . & . & . & 0 \\ A_{n1} & A_{n2} & . & A_{nn} \end{bmatrix} \quad (5)$$

Such a matrix defines *a decoupled design* (conception). When the design matrix is triangular, the independence of functional requirements is ensured by setting the *DPs* in a certain order. Equation 5 shows a decoupled design. DP_1 is the first to set, followed by DP_2 and so on. If the design matrix cannot be reduced to a triangular matrix, the design is *coupled.*

Therefore, the design should have a diagonal design matrix or a triangular matrix.

The following example shows a decoupled design, where each X is a non-zero element:

$$\begin{Bmatrix} FR_1 \\ FR_2 \\ FR_3 \end{Bmatrix} = \begin{bmatrix} X & & \\ X & X & \\ X & X & X \end{bmatrix} \cdot \begin{Bmatrix} DP_1 \\ DP_2 \\ DP_3 \end{Bmatrix} \quad (6)$$

(c) *A coupled design matrix* contains non-zero elements above and below the diagonal that columns and row changing cannot solve:

$$[A] = \begin{bmatrix} A_{11} & A_{12} & . & A_{1n} \\ A_{21} & A_{22} & . & A_{2n} \\ . & . & . & . \\ A_{n1} & A_{n2} & . & A_{nn} \end{bmatrix} \quad (7)$$

The number of *FR*s (functional requirements) equals the number of *DPs* (design parameters) necessary to achieve an ideal design. Thus, the design matrix [*A*] is square. The independence axiom is met for uncoupled or decoupled designs. Uncoupled designs ensure *an ideal design.*

However, the number of *FRs* may differ from the number of *DPs*. In this case, the following design categories are:

(1) If the number of *FRs* exceeds the number of *DPs*, the design equation is as in the following example:

$$\begin{Bmatrix} FR_1 \\ FR_2 \\ FR_3 \end{Bmatrix} = \begin{bmatrix} X & 0 \\ 0 & X \\ A_{31} & A_{32} \end{bmatrix} \begin{Bmatrix} DP_1 \\ DP_2 \end{Bmatrix} \quad (8)$$

It shows a *Coupled design* or some *FRs* cannot be fulfilled. The previous relation shows FR_3 cannot be met if A_{31} and A_{32} are zero. However, if at least one of the two elements has non-zero values, we are dealing with *a coupled design* [15].

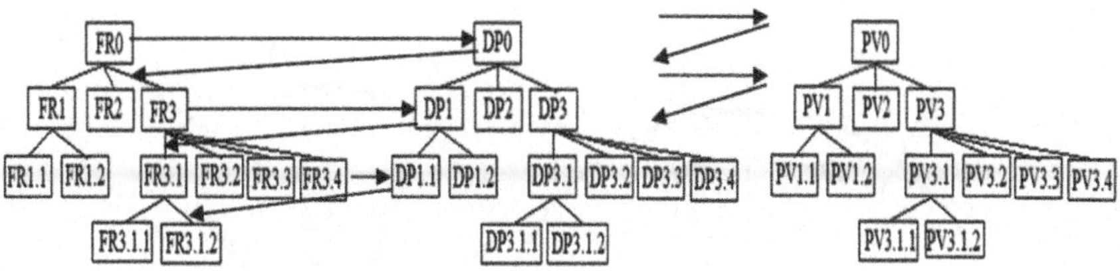

Fig. 4 Hierarchical decomposition of *FRs* functional requirements, design parameters *DPs* and *PVs* process variables by zigzagging

As mentioned before, *if the number of functional requirements is equal to the number of design parameters* (and when we are dealing with a diagonal matrix or a triangular matrix) *the axiom of independence is met.* We deal with an *ideal* or a *decoupled design.*

(2) *A design is a Redundant design* if the number of *FRs* is less than the number of *DPs*. The relation illustrates an example:

$$\begin{Bmatrix} FR_1 \\ FR_2 \end{Bmatrix} = \begin{bmatrix} A_{11} & 0 & A_{13} & A_{14} \\ 0 & A_{22} & A_{23} & A_{24} \end{bmatrix} \begin{Bmatrix} DP_1 \\ DP_2 \\ DP_3 \\ DP_4 \end{Bmatrix}. \tag{9}$$

The zigzag decomposition takes place between domains and across all domains, as depicted in Fig. 4. From FR_0 the design goes to DP_0 and eventually to PV_0. Thus FR_1, FR_2, and FR_3, are images from DP_0 that exhaustively fulfills FR_0. Each *FR* allows reaching a *DP*, DP_1, DP_2, and DP_3. The process ends at the leaves DPs. A leaf is a single part that can be manufactured or an artifact accessed in the market.

3 Using the First Axiom of Axiomatic Design

The following example briefly presents how to use the first axiom of Axiomatic Design. The first axiom is widely used, while the second has fewer applications due to the challenge of achieving good designs. For this reason, this introductory chapter of AD will only present an application of the first axiom.

The example is the design of a polishing device. The polish device rotates brushes with abrasive blades (Fig. 5) that allows achieving a good adaptation of the blades to the shape of the helical groove. The high-speed rotating brush has a portable device with a shaft to fasten radial abrasive blades made of cardboard or textile. These blades incorporate abrasive grains on one or both of their active surfaces.

Such a device is necessary for polishing the surface of a helical groove located on an outer cylindrical surface. Polishing should ensure a low roughness at the groove surface. The surface to polish is part of the equipment that moves food materials in

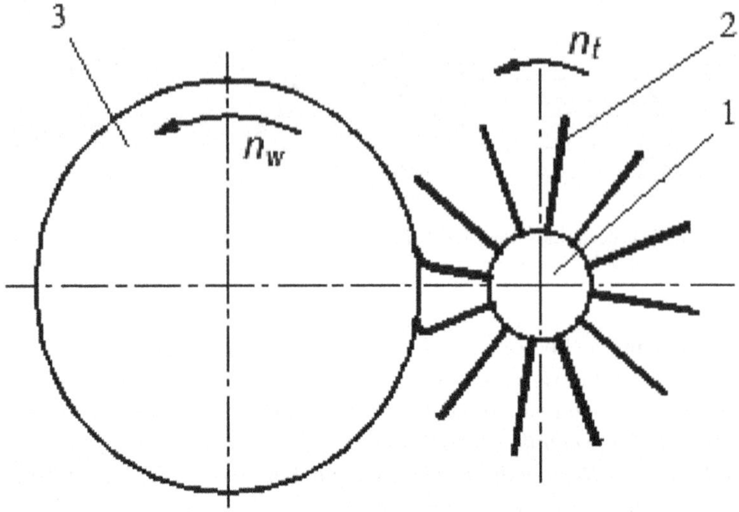

Fig. 5 Schematic representation of the polishing process using rotating brushes with abrasive blades: (1) the shaft of the rotating disk driven in a rotational movement with speed n_t; (2) abrasive blade; and (3) workpiece in rotating motion; n_w—the rotational movement of the workpiece (Adapted from [16])

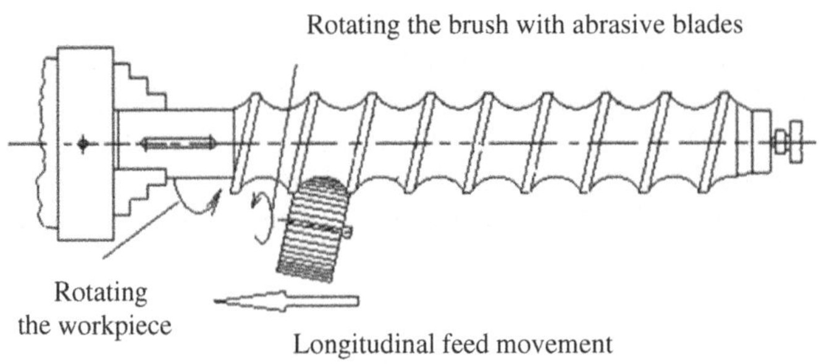

Fig. 6 Machining scheme valid in case of polishing a helical groove using the rotating brush with abrasive blades (Adapted from [17])

a cylinder. Low roughness ensures not to retain food debris on the surface. Figure 6 shows the process, the rotation of the workpiece, and the longitudinal movement of a rotating brush.

The device is part of experimental research activity of the influence exerted by the parameters specific to the polishing process (considered here as input elements in the system related to the polishing process) on the values of parameters of technological interest (process output parameters). The input parameters are the dimension and material of the small abrasive grains, the rotating speed of the brush and workpiece. Finally, it is taken into account the vertical and horizontal angles of the brush axis.

To further develop the experimental research concerning the polishing process, the following customer requirements were formulated:

CN_1: The device must be able to be used on an existing machine tools in a small mechanical workshop;
CN_2: The device must provide conditions for changing polishing parameters.

Notice that the Customer's Needs are about a device, not the polishing process. Therefore, the polishing input parameters regarding the abrasive material are not considered. It makes the main functional requirement (zero-order) to be:

FR_0: Research on the polishing process of a helical groove.
DP_0: Polishing device at a workpiece location.

The zero-level functional requirement may be used to design a simple DP product. However, if the product is not available, it is necessary to decompose the functional requirements into lower levels (levels 1, 2, 3, etc.). In the situation under analysis, the design is decomposed until the second level. The first level decomposition is highlighted in Table 1, showing the (FR_i) and (DP_i).

Figure 7 presents the proposed design for the polishing device. It is found that for polishing the existing helical groove in the workpiece located and clamped in the universal chuck and the live center, a portable drilling machine is used. This portable drilling machine is clamped through a bracelet-type device on the disc, which can be rotated and fixed in a certain angular position using some nuts. The respective nuts are screwed onto threaded rods, which are secured to the fixed disc. The threaded rods pass through a groove in the form of an arc of a circle existing in the disc, which can be rotated and thus immobilized in the desired position.

The detailed description of *FRs* and *DPs* at the second level of decomposition is described in Table 2.

The device can be clamped in one of the four locations of a lathe ordinary tool holder due to its endowment with a part that has a parallelepiped-shaped step. The abrasive blade brush materializes the polishing process. The guide of the tool slide can be rotated and fixed at an angle whose value is determined by taking into account the angle of inclination of the helical groove in the workpiece. The immobilization

Table 1 FRs and DPs on the first level of decomposition

Functional requirements	Design parameters
*FR*1: Rotate the brush with abrasive blades at different rotational speeds	*DP*1: Portable drilling machine
*FR*2: Rotate the workpiece with different rotation speeds	*DP*2: Universal lathe
*FR*3: Adjust the angle of the vertical axis of the brush	*DP*3: Mechanical subsystem for rotating and fixing the brush at a certain vertical angle
*FR*4: Adjust the angle of the horizontal axis of the brush	*DP*4: Mechanical subsystem for rotating and fixing the brush at a certain horizontal position

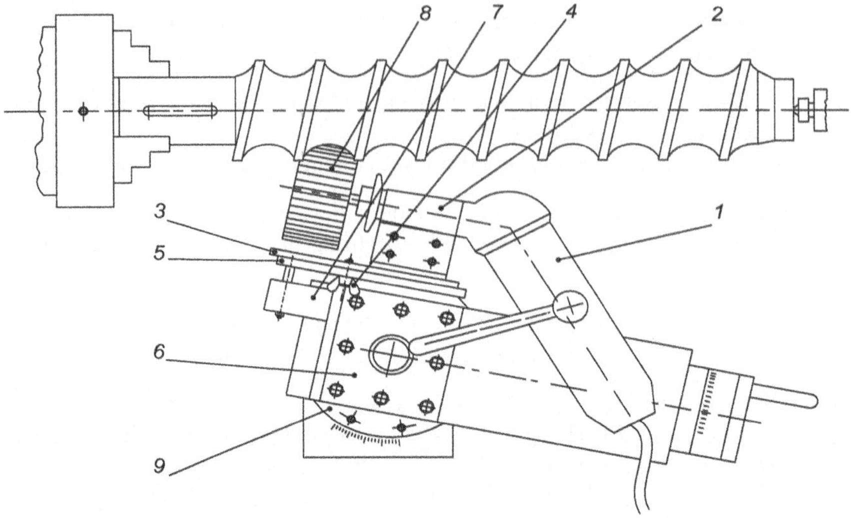

Fig. 7 Device for polishing the surface of a helical groove placed on a cylindrical surface using rotating brushes with abrasive blades: (1) portable drilling machine; (2) bracelet device for clamping; (3) disc; (4) nuts; (5) fixed disc; (6) lathe tool holder; (7) parallelipipedic-shaped part; (8) abrasive blade brush; (9) guide of the tool slide (Adapted from [17])

of the guide in the desired angular position is done using nuts. All these components are parts of the usual equipment of the universal lathe.

The axiomatic design has been used to analyze older projects and identify ways to improve them. Sundar et al. [18] expressed an interesting point of view. They set out to investigate how the human ear meets the specific requirements of axiomatic design.

4 The AD Second Axiom

According to AD, the design is a decomposition process from the highest levels to the sleeve level. At each level of decomposition, the first and Second Axiom applies.

In the previous section, we saw ways to classify the designs. The design to choose might be uncoupled or decoupled. A coupled design is a poor design. From all acceptable designs, the one with the highest probability of success must be the one to choose.

This section presents some of the most common methods to evaluate the probability of success. It shows the definition of information from the probability density function (pdf), the evaluation with fuzzy logic, and the Dempster-Shaper method. Moreover, it gives place to Delphi methods to evaluate information.

Table 2 Matrix comprising *FRs* functional requirements and *DPs* design parameters in the case of a polishing device using rotating brushes with abrasive blades

Line no. 1	Design parameters			Design parameters DP_0: *Polishing device*							
2				*DPs* design parameters of the first order							
3				*DP*1: Portable drilling machine		*DP*2: Lathe		*DP*3: Mechanical subsystem for rotating and fixing the brush at a certain vertical angle		*DP*4: Mechanical subsystem for rotating and fixing the brush at a certain horizontal position	
4				Second order *DPs* design parameters							
5	Functional requirements			*DP*1.1: Electric motor of the portable drilling machine	*DP*1.2: Mechanical subassembly of the portable drilling machine	*DP*2.1: Lathe electric motor	*DP*2.2: Lathe gear box	*DP*3.1: Slide guide of the rotary tool holder that supports a vertical axis	*DP3.2:* Nuts for fixing in the position obtained by rotating around a vertical axis of the slide guide of the tool holder	*DP*4.1: Disc holder that can be rotated relative to a fixed disc	*DP*4.2: Nuts for securing the brush holder disc with the fixed disc
6 Column no. 1	2	3	4	5		6		7	8	9	10
7	Zero order functional requirement	First order *FRi* functional requirements	Second-order functional requirements	Highlighting the *DPi* design parameters corresponding to each *FRi* functional requirement							

(continued)

Table 2 (continued)

8		*FR*1: Rotate the brush with abrasive blades at different rotational speeds	*FR*1.1: Rotate the brush with the abrasive blades	X							
9			*FR*1.2: Change the rotation speed of the brush		X						
10	*FR*0: Research on the polishing process of a helical groove	*FR*2: Rotate the workpiece with different rotation speeds	*FR*2.1: Rotate the workpiece			X					
11			*FR*2.2: Change the rotation speed of the workpiece				X				
12		*FR*3: Adjust the angle of the vertical axis of the brush	*FR*3.1: Put the axis of the brush with abrasive blades relative to a vertical axis					X			

(continued)

Table 2 (continued)

13			*FR*3.2 Fix in the axis of the brush with abrasive blades in a certain position to a vertical axis						X		
14		*FR*4: Adjust the angle of the horizontal axis of the brush	*FR*4.1: Put the axis of the brush with abrasive blades relative to a horizontal axis							X	
15			*FR*4.2: Fix the axis of the brush with abrasive blades in a certain position to a horizontal axis								X

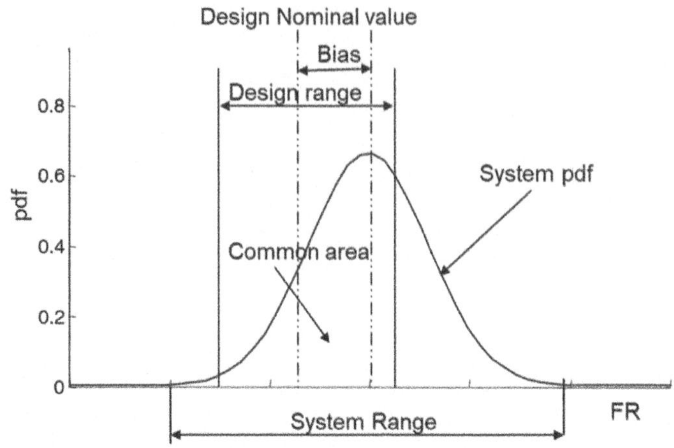

Fig. 8 Probability density function for a design with a single FR

The Second Axiom of AD states: minimize the information content of a design. The information is a measure of the probability of success so that it allows to choose the design with the highest probability of success.

Figure 8 shows the pdf of a single FR, the system pdf, the design range, and the common area. The common area is the intersection between the system area and the design boundaries defined for the FR.

The FR is accomplished if the working system is within the Design range. Figure 8 also shows the nominal value of the design and the bias to the system probability density function (pdf). The more extensive the design range, the higher the probability of success of the design. Notice that the computation occurs in the Functional Domain, not the Physical Domain.

The probability of success is the relation between the common area and the system area (Eq. 10). The system area is unitary when using a pdf.

$$p = \frac{Common\ Area}{System\ Area} \tag{10}$$

According to the Shannon equation, the information content, I, expressed in bits for a single FR is defined in Eq. 11. The lower the information, the higher the probability of success.

$$I = \log_2(1/p) = -\log_2(p) \tag{11}$$

In many designs, the system pdf is unknown. If the designer achieves a mean and standard deviation, a normal distribution is defined, which allows estimating the probability. For example, the average power of an equipment and its standard deviation can come from the heat needs and schedules.

In other applications, there is no knowledge about the pdf. However, the system range is known, making it possible to compute the information using uniform distributions. Uniform distributions commonly allow the evaluation of the pdf for mechanical tolerances. It can be used to calculate the information of two or three FRs for any design, uncoupled, decoupled, or coupled.

Fuzzy logic has been used to evaluate the system pdf by using the system membership. The System membership function can be a triangular, trapezoidal, sigmoid function, or any user-defined or computed. The system range is usually known, and the function shape can be selected from the designer's experience. Moreover, FR may arise from fuzzy algebra if a mathematic expression defines the FR. Each parameter of the mathematic expression has a membership function.

Figure 9 shows an example of a system membership computed from a mathematic expression.

The Figure 9 depicts a system membership function computed from the product of fuzzy numbers. The design membership is a trapezoidal function in the example, other than a crisp design range.

The system membership area is not necessarily unitary in the fuzzy sets theory. Therefore, Eq. 11 applies to pdf applications as well as fuzzy applications.

As the reader already read, a design with more than one FR system can be uncoupled, decoupled, or coupled. As already mentioned, from the available designs, the one with the minor information content must be the design to choose.

The matrix shape is essential to define the information content of the design.

If the design is uncoupled, all FRs adjust independently. The join probability P of all independent events is the product of all probabilities. Therefore, the information content is the sum of the information of the FRs, as depicted in Eq. 12.

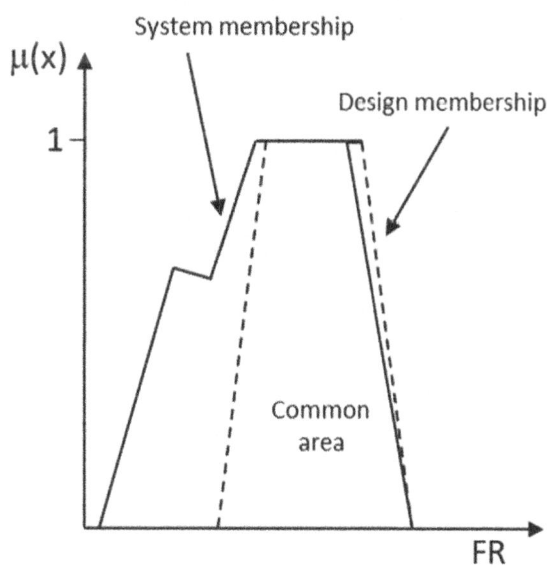

Fig. 9 Membership function for a single FR

$$I = -\log_2(P) = -\log(\prod_i p_i) = \sum_i (-\log_2(p_i)) = \sum_i I_i \quad (12)$$

Each probability p_i is achieved by knowing the system pdf of a system membership function.

If the design is decoupled, there is a sequence to tune the FRs. Referring to Eq. 6, FR_1 is the first FR to adjust, then FR_2, and finally FR_3. In other words, FR_2 is subjected to FR_1, and FR_3 to the occurrence of FR_2 and FR_1. The probability problem turns into a Bayesian probability. For decoupled designs, the probability is according to Eq. 13:

$$P = p(FR_1) \cdot p(FR_2|FR_1) \cdot p(FR_3|FR_1, FR_2) \quad (13)$$

In sequential applications, the probability of fulfilling an FR is in the context of all former FRs. In a failure evaluation, an FR_2 is a function affected by a former letdown of FR_1.

In many engineering situations, the picture is well described by fuzzy logic. Experts can estimate ranges of values of an event subject to the occurrence of previous events. Moreover, they can define a type of fuzzy function from their knowledge of the problem. Triangular functions are the most used ones. As a result, the probability of the design is the product of all fuzzy events in a sequence.

Similar reasoning applies using the Dempster-Shafer Theory (DST) to estimate the information of a decoupled design.

DST uses belief, a measure of what we know, and plausibility, the measure of what we can know about the event or scenario. The design can be defined in different uncertain scenarios at the same time. Therefore, DST is most attractive for the higher decomposition levels of any design when knowledge about the complete system is scarce.

In DST, the events have a frame of discernment Θ with n mutually exclusive and exhaustive singletons. Suppose the frame of discernment has three singletons $\Theta = \{f_1, f_2, f_3\}$. It can correspond to a known scenario f_1, improved scenario f_2, and unknown scenario f_3. The set of all subsets Θ is the power set P with 2^n sets.

For three singletons,

$P = \{\emptyset, \{f_1\}, \{f_2\}, \{f_3\}, \{f_1, f_2\}, \{f_1, f_3\}, \{f_2, f_3\}, \{f_1, f_2, f_3\}\}$.

Let A be any set of sets of P. It represents a proposition $A \in P$. The example can be f_1, f_2, f_1, and f_2, and none of them.

The mass probability $m(A_i)$ is the basic probability assignment (*bpa*). The focal elements are the sets with non-zero mass. The *bpa* can be used to compute the information content. Likewise, the *pdf*, *bpa* has the properties shown by Eq. 14:

$$m(\emptyset) = 0, \qquad \sum\nolimits_{A_i \in P} m(A_i) = 1 \quad (14)$$

Mass probability maps the values of a power set, not from the x values of a universe of discourse. The belief in a set A_i, is the sum of mass evidence in the subsets B such that $B \subseteq A_i$ (Eq. 15):

$$bel(A_i) = \sum_{B \subseteq A} m(B) \quad (15)$$

Equation 16 shows the plausibility $pl(A_i)$. It is the evidence in all subsets of A_i that intersect B.

$$pl(A_i) = \sum_{B \cap A \neq \emptyset} m(B) \quad (16)$$

Bl(A) and *pl(A)* allow computing the information content. Therefore, it makes *bl(A)* and *pl(A)* the lower and upper bounds of the probability of fulfilling an FR.

Finally, despite the lack of scientific foundation of the Delfi models, it is worth presenting it because of their broad application. Delphi models are an iterative process of Estimate-Talk-Estimate. A panel of experts compares the probability of success of different designs. Then they discuss the results and compare them again. The process is supposed to converge to a decision. The group must have all the necessary skills to make a proper decision. Delphi methods can be applied to evaluate the probability of success of different designs. Moreover, a panel of experts can decide between two coupled designs, case no uncoupled design is available.

5 Comparison with Other Design Methods

This section analyzes common concepts, values, practices, and assumptions between AD and other design methodologies. It explores different methods to create a design framework together with AD. We will call framing how to merge concepts, values, and practices from several design methods to create frameworks that ensure a greater capability to create value.

5.1 Context of Comparison with Other Design Methods

Design Theory and Methodology (DTM) is a field of design research that deals with the study of design principles, knowledge, procedures, and practices. We could say that the objective of the DTM is instead focused on the way it is designed (principles, practices, activities) than on what is designed (products or services).

The ways in which scientific knowledge evolves are complex and different. The early stages are based on observations, experiments, events, intuition, imagination, and the gain is a collection of facts. The facts are validated by testing strategies in the context of hypotheses. Not infrequently this road encounters many difficulties.

We could consider that scientific knowledge starts from facts interpreted in a field of hypotheses to reach laws.

This is also the case with DTM which begins with a collection of individual design cases and evolves into more abstract and general forms intending to become a general (universal) or abstract theory about design.

There is a wide variety of DTMs. The elaboration of a list of DTM may risk not including all methodologies. A comprehensive list of twenty-three DTMs is summarized in [19].

Over time, several classifications of DTMs have been attempted. Tomiyama proposed one of the classifications that seemed appropriate to this approach based on GDM (General Design Theory) founded by Yoshikawa [19].

According to GDP, knowledge can be mathematically formalized, and for this three axioms are proposed:

- The first axiom ("axiom of recognition") according to which any entity can be identified and modeled by attributes;
- The second axiom ("axiom of correspondence") establishes a one-to-one cardinality between the set of entities and the set of entity concepts;
- The third axiom ("axiom of operation") is according to which the set of entity concepts and the set of abstract concepts form a topological space.

Mathematically, theorems may be derived from these axioms to explain the design process.

From a GDT perspective, the design process is a mapping of the functional space to the attribute space, defined on the set of conceptual entities, as exemplified in the GDT proposed by Tomiyama [19]. Three different categories of using DTMs in design activity have been identified, such as [19]:

- *Using DTM to generate a new design solution* (based on creativity, modification, adaptation, combination, systematic approach, etc.);
- *Using DTM for the development of adjacent functionalities* (QFD, AD, DfX, Taguchi, etc.);
- *Use of DTM for design knowledge management* (design knowledge management, concurrent engineering, etc.)

These categories are not mutually exclusive, and some design methodology overlaps two or more areas.

However, Axiomatic Design appears in categories together with Taguchi method, QFD, DfX, FMEA, analysis technique, optimization technique, and genetic algorithm. Therefore, the Axiomatic Design can relate to other methods in the same category (compare what is comparable).

In the following subsections, we will analyze/present comparisons of the Axiomatic Design methodology with the Taguchi, QFD, and TRIZ methodologies.

5.2 Comparison AD and Taguchi Method

The literature in the field [19–22] states that the method was first proposed by Genichi Taguchi in the 1950s in Japan at NTT (Nippon Telephone and Telegram Coop.). It was also stated that the Japanese industry accepted it well. In the 1980s the method was introduced in the US and then in Europe.

Robust parameter design is a method of systematic application of DoE (Design of Experiments) to optimize projects by improving their transfer functions.

The central concept of the method is the sensitivity of a design to the uncontrolled factors encountered in production as well as in use.

According to the method, the loss of quality during the life cycle is assimilated with the deviation from the desired performance, and a good design minimizes the loss of quality. Such a model is considered as robust as it is less sensitive to noise.

The concepts of Taguchi theory could be summarized as follows:

1. The primordiality of quality in the design phase and not of inspection;
2. Immunity of the product to uncontrolled environmental factors;
3. Evaluate the quality costs of the entire system by measuring the deviation from the standard.

In order to achieve the desired quality of the product through design, the methodology recommends a three-phase approach:

1. Systems design—identifying appropriate work levels for design factors.
2. Parameter design—determination of factor levels based on the condition that the influence of uncontrolled factors (noise factors) produces minimal variation in system performance.
3. Tolerance design—focusing attention on the tolerance of the factors determined in the previous stage, factors that have a significant influence on the product.

The Taguchi method uses the concept of the loss function to define the evaluation of product quality, a function that expresses the loss in use of a product due to variations in product function and other losses (secondary costs). Losses are estimated by mean square deviation from the target value and the smaller they are the more robust the design is.

The use of Taguchi methods for the application of design axioms is an approach in which the author argues that the engineering analysis methods developed by Taguchi are consistent with the two axioms of axiomatic design set forth by Suh [23]. The proposed framework, especially when the number of requirements is very high, starts with AD by organizing the problem in terms of functional requirements. Each functional requirement requires a controlling factor that can be determined using the Taguchi method and the experimental design. The author concludes that AD's language is different from that of Taguchi methods, although the principles are the same: independence of functional requirements and minimization of design information content. There are examples where AD is combined with several tools (the seven quality control tools) and the design of experiments [24]. The link between the complexity and robustness of a system is a topic debated in the literature. AD and Taguchi methods are inevitably taken into account when discussing this topic. Gohler presents a quantitative approach to relate robustness and complexity using a model-based probabilistic [25]. Authors define complexity through the degree of coupling (directly related to axiom 1 of AD) and the level of contradiction between functional requirements. The problem that the authors set out to clarify is whether there is an association between the degree of coupling and the level of contradiction of a design on the one hand and its robustness on the other.

There are many and varied approaches to developing frameworks for design. An example is a framework for robust design and Variation Management Framework (VMF) by combining central models to Robust Design, and the Domains of Axiomatic Design [26]. The authors concluded that VMF has proven to be a valuable framework to communicate robust design and variation to engineering and senior management levels.

An example of sustainable product development by integrating Robust Design criteria and Axiomatic Design principles is presented by [27]. The authors propose a design framework in four steps. Once the customer requirements are selected, the axiomatic design process proceeds.

Product design frameworks ensure quality in the conceptual stages instead of quality inspection, where the quality of a product remains undetermined until the product is built and tested [28]. The author proposes a methodology for integrating design for quality in modular product design by considering the underlying principles of axiomatic design and robust design along with the product's perceived quality. For evaluating the modular architecture, metrics are defined. Each module (Membership functions) is evaluated based on robustness and compliance with the axiomatic design principles.

Oh [29] equates the relationship between functional requirements (FRs) and design parameters (DPs) as a transfer function. Suppose a specific value of a DP determines that the corresponding FR reaches a particular target. In that case, the variation of this design parameter will determine a variation of the corresponding functional requirement around a value. A Taylor series expansion can approximate this variation of the functional requirement. The transfer function is developed using two matrices in the relationship between functional requirements and design parameters: one that reflects independence (Axiom 1), called the matrix [A], and a second that reflects the informational content (Axiom 2) named matrix [B]. Through this approach, the mathematical treatment between Axiomatic Design and Robust Design was extended.

In a mixed approach for robust design integrating the Taguchi method in Axiomatic Design [30], the authors consider that the main difficulty in checking the second axiom of AD theory is the identification of relations among FRs and DPs. In the case of an uncoupled problem, each DP regards only one FR. The most robust solution is the one with the lower "information" or higher probability of success. For uncoupled designs, information is the sum of the effects from each relationship. The problem is complicated in the case when it is modeled as coupled, where more DPs influence the same FR. The authors consider that the relationship law between

FRs and DPs can be identified by working on a physical prototype of the product. During concept design and early embodiment, the second axiom cannot be applied. Taguchi method can supply this kind of information since the preliminary phase of an embodiment when first product architectures appear, and suggestions about dimensioning and material choice for each component can guide the designer toward a deeper knowledge of the solution he/she is pursuing. The first phase of the design process take place in two steps:

- check by means of the first Axiom of Axiomatic Design if the solution is good;
- measure by means of Taguchi Method the level of robustness of the solution.

Several design methodologies such as axiomatic design, robust design, and the theory of inventive problem-solving have been integrated with the functional prioritization framework provided by reliability-centered maintenance to develop a new conceptual design methodology [31]. To propose a framework that encompasses the four methodologies, the authors compared their features.

Yihai [32] propose a design framework that combines AD, TRIZ, and Taguchi. AD provides an analysis function to find latent contradictions, and TRIZ is used to solve specific contradictions by contradiction matrix and solving principles. The Taguchi method is used for parameter optimization—parameter design and tolerance design.

5.3 Comparisons Axiomatic Design and TRIZ

The name of the method "TRIZ" comes from the Russian title of the book "Theory of Inventive Problem Solving":—Theory Resheniya Izobretatel'skih Zadach (TRIZ), being an acronym [33, 34]. TRIZ is the work of Genrich Altshuller and consists of the formulation of a number of generally applicable inventive principles that resulted from the analysis of forty thousand patents.

TRIZ is an engineering problem-solving toolkit that systematically uses known solutions to solve future problems. TRIZ is used for each stage of problem-solving by preparation for problem-solving, problem-solving, and solution selection and development [34]. The TRIZ methodology is oriented toward an ideal end result (Ideality) and leads the user to inventive solutions, rejecting compromises as a possible result. Identifying contradictions and applying the principles of solving them involves a systematic direction of solving problems.

There are many ways AD and TRIZ are combined in design methodologies.

Targeting industry best practices, Borgianni and Matt developed a study that looked at the application of AD and TRIZ methods and developed a classification of AD and TRIZ applications in different industries [35]. The study was based on articles published in 2014 and 2015 on the application of TRIZ, and Axiomatic Design reported in Scopus-indexed. From the trends revealed by applying AD and TRIZ methods, we can deduce the decrease of their mutual/synergistic implementation. The authors also consider that the sequential use of the two methods, AD (to analyze the problem) and TRIZ (to resolve circumstantial contradictions), is ineffective and suggest ways to exploit the opportunity to build a new framework capable of addressing issues related to complex systems.

Understanding TRIZ through the review of top-cited publications [36] is a paper reviewing the literature, including the top 102 indexed publications concerning TRIZ, according to the number of citations received. The description of the TRIZ application fields and their use in different work frameworks are organized in clusters. Clusters 5–6 refer to TRIZ for ideation and conceptual design used as a stand-alone methodology and combined with other techniques (Case-Based Reasoning, Axiomatic Design, etc.).

Regarding the relationship between AD and TRIZ, the authors opine that there are two distinctive features of the AD paradigm that can be seen as complementary aspects of the TRIZ application domain:

- The first distinctive feature is that AD focuses on functional requirements and relates them to physical requirements.
- The second distinctive feature consists of the existence of the two axioms that help distinguish good from bad designs. The design classification allows the hypothesis of a framework integrating AD and TRIZ: AD is firstly used to formulate technical requirements (problem setup), then TRIZ is entrusted to the invention, and solutions are evaluated from axioms perspective.

A methodology to conceptually design firmware that will help bridge the gap between software and hardware conceptual design is presented in [37]. The proposed framework integrates UML (Unified Modeling Language), AD, and TRIZ. A conversion method between the axiomatic design matrix and the widely used UML sequence diagram was developed. DPs of the design matrix are defined as the objects in the UML sequence diagram, and FRs of the design matrix are generated by merging FMs depending on their flow of information in the sequence diagram. According to the authors, the methodology developed helps bridge the gap between axiomatic design theory and software design and creates the possibility of improving the full integrated system (hardware and software) simultaneously.

A framework for solving problems using synergistically TRIZ and AD is proposed in [38]. The authors consider that by applying AD and TRIZ in a framework, the strengths of both methodologies are capitalized. First, AD is applied to analyze the problem and break down the main problem into a hierarchy of problems, and then TRIZ is applied to generate innovative solutions to the problems in the previous hierarchy. In this way, the framework formed by AD and TRIZ uses synergistically the capacity of detailed analysis of AD with the innovative process of generating ideas of TRIZ.

There are conceptual design approaches in which AD is used in a TRIZ framework [39]. According to the authors, the strengths of the methods are:

For TRIZ—problem identification (contradiction) and concept generation.
For AD—problem identification (coupling) and formulation steps.

The authors consider that the axiom of independence of AD can be used to narrow down the list of possible standard solutions generated by the existence of a physical contradiction (according to the TRIZ principles).

In [40], the authors summarize the possible relations between Axiomatic Design rules and TRIZ problem-solving tools. Seven corollaries that serve as the design rules are directly derived from two axioms, so comparing these "lower level design rules" with TRIZ tools is useful for understanding these two methodologies.

5.4 Comparisons Axiomatic Design and QFD

Quality Function Deployment, or QFD, is a method and a set of tools for product development. QFD is used to effectively define customer requirements and convert them into detailed engineering specifications and plans to produce the products that fulfill those requirements. QFD was founded in the 1960s by Mizuno and Akao in Japan [19].

QFD process may be different, depending on the types of products, such as improvements of existing products, innovative new products, mass production products, order-made products, etc. QFD is implemented iteratively, and each iteration consists of mapping some quality elements into other quality elements by using a matrix formulation (called the House of Quality). Iterations of QFD implementation are presented as follows:

- Product planning: identify customer requirements; translate VOC (voice of customers) into design specifications; prioritize requirements; evaluate the competition.
- Product design: generate design ideas or concepts; translate the outputs of the product planning phase into individual part details, identify product risks, define the product specifications.
- Process planning: defines the product development process; establishes process controls; creates a manufacturing process flowchart and process parameters.
- Process control (production planning): define the production requirements for each component/operation; establish inspection and test methods; define performance indicators to monitor the production process.

The basic design tool of quality function deployment is the house of quality. This tool allows the identification and clarification of the client's requirements (What's), identifies the importance of these requirements, identifies the engineering characteristics relevant to these requirements (How's), and the correlation of the two allows the assignment of objectives and priorities. House of quality summarizes customer requirements, weights, and correlation matrix of customer requirements and technical specifications using a matrix form.

There are a variety of attempts to create frameworks by identifying concepts, best practices, common values between AD and QFD (even with other methods). Some of these attempts are presented below.

A literature review study on comparing and integrating AD with QFD as a design method is presented in [41]. The authors developed a comparative analysis between the two methods based on this study. The authors conclude that the QFD method is used to identify the problem that occurs early in the design, and Axiomatic Design is more suitable for product development with high quality. By integrating both methods, the axiomatic design is used to analyze the systematic changes of customer requirements into design parameters, functional requirements, and process variables from the house of quality.

Attempts to develop frameworks for design sometimes extend to combining several methods organized into different flows [42]. The authors present a method of integrating QFD, AD, and benchmarking methods (BM) to conduct the searching process of design solutions.

The proposed method consists of three processes: applying QFD to map customer requirements (CRs) into function requirements (FRs); mapping FRs into design parameters (DPs) by using AD; and applying the benchmarking method (BM) to search for optimal design specifications by the comparative analysis and concept combination from benchmark products.

5.5 Discussion

All above-analyzed design methods are part of DTMs and could be used to choose or optimize products or processes and solve problems in general. Various literature review articles signal the organization of design methods in different frameworks. Each framework proposal is accompanied by case studies, arguments that support the organization in the proposed way. However, there are rare proposals to validate different methods or frameworks.

This subchapter considers four design methods and philosophies that are used nowadays in the industry.

Some considerations about comparison criteria, differences, and similarities are synthesized in Table 3. Table 4 highlights the main issue. Notable differences can be identified between methods but also common or similar things.

AD is the only method that concentrates on obtaining an ideal design. It is possible to propose a new design or to analyze an old design, and the axioms are the scientific bases used for this goal. The concept of mapping or zigzagging process belongs only to AD, and it allows to avoid the relations between the function of the product, processes, or systems.

TRIZ's concept of solving contradictions is similar to the idea of independence from the first axiom of AD. Only AD and TRIZ handling with the problem of functional coupling.

QFD concentrates on satisfying the customer needs, and the importance is given to the VOC is similar to the importance of choosing the FRs in AD because a design solution cannot be better than its FRs. However, QFD creates coupled designs in the

Table 3 Comparison between AD, Taguchi Method, TRIZ, QFD

	AD	Taguchi Method	TRIZ	QFD
Main objective	• Ideal design	• Minimize variability • Quality loss function	• Solve technical problems	• Set targets for the technical attributes of a product • Understand the importance of each/hierarchization
Main output	• New design that satisfy all customer needs • Improving of an existing design	• Robustness of the quality characteristics	• Elimination of contradictions • Technical improvement	• Satisfaction of customer needs
How to improve the design	• Trying to eliminate the relationship between functions • All functions need to be fulfilled, so there are no function weights	• Optimize/minimize the quality loss function • Take into consideration the cost	• Incorporate quality and reliability in the design stage	• Increasing the parameters that have a most important impact in some perspective
Ease of use/application	• Medium difficulty • Not widely spread	• Medium difficulty	• Relatively widely spread	• Relatively easy • widely spread
Main strength/ advantages	• Identify the best design easily • Suitable for the decision making of product development with high quality	• Transform the variation from nominal value taking into consideration with financial depiction	• Focus its studies on inventive problem-solving • Generating creative design solutions	• It is a customer driven process
Disadvantages	• The creative process comes from applying the two axioms but is not straightforward	• Define tolerances of a design instead of changing the design to allow higher tolerances	• Lack of formalization	• Sometimes, it does not use meaningful information • Creates a coupled design • Prioritize functions

Table 4 Main issues regarding the comparison between AD and Taguchi method, TRIZ, QFD

Differences/similarities	Taguchi method	TRIZ	QFD
AD	• Taguchi takes into consideration some components of the cost • Taguchi method does not implie a zig-zagging process • Concentrate on tolerance specification, rather than design change to allow higher tolerance	• TRIZ contradiction concept is similar to the functional coupling in AD. Overcoming contradiction means the removal of functional coupling • TRIZ has no method to identify couplings • AD theory states the general rules of engineering design to help innovation; TRIZ methodology concentrates on inventive problem-solving techniques for coupling problems	• Produce a coupled design • VOC ≈ CNs • prioritize functions rather than using the minimum number of functions with no weights

vast majority of applications. Moreover, QFD prioritizes functions, and AD aim to define the minimum number of functions necessary to fulfill the Customer Needs.

The Taguchi method can be used efficaciously in combination with AD. After obtaining a good design, the level of its robustness could be measured. Taguchi's main objective is to minimize the quality loss function. This is similar to the AD second axiom, which states minimizing the information content of a design that represents choosing the design with the highest probability of success. However, the approach is different—Taguchi method aims to reduce the deviation from a target by reducing tolerances; AD aims to define tolerances as large as possible and thus define a system within the range.

Taguchi is the only method discussed in this chapter that explicitly considers the cost to improve the design. Cost can be a constrain in AD or defined as a function.

6 AD Application

As can be seen from the above, the first applications of axiomatic design aimed at developing the design of manufacturing technologies. Axiomatic design can be used efficiently in the case of constructive design activities, and so far, quite a lot of applications have been identified in this area.

However, there are many other areas without apparent connections with manufacturing technologies' design or constructive design. Efficient results have been obtained by applying axiomatic design to them.

There are various ways of identifying sectors of human activity in which design or analysis activities have so far been undertaken based on axiomatic design principles. Economic and social life fields have applied AD and are briefly presented below, taking into account the classification of the sectors of activity proposed by Borgianni and Matt in 1996 [35].

Design of vehicles and vehicle components. Naddeo addressed the issue of designing a car platform usable in the case of an electric rear-wheel-drive vehicle. He also used a fuzzy approach to this problem, reaching optimized solutions for battery placement and car platform crossbars as chassis components [43].

Tate et al. used AD to develop competitive cars in the Texas Tech University Eco-CAR program [44]. The AD use led to the decision to use a two-mode hybrid architecture. The correctness of the selected solution was verified by simulation.

Materials processing. Kazmer used axiomatic design principles to improve the control of the injection molding process [45]. Thus, he found that multi-cavity pressure control contributes to a spatial decoupling that increases the number of degrees of freedom that define the quality characteristics. Also, the dynamic temperature control ensures a temporary decoupling of the injection and solidification stages, which facilitates the identification of ways to increase the performance of the injection molding process. Esther Richards used the axiomatic design to design and materialize an apparatus usable in evaluating the gas solubility in polymers [46].

Civil engineering. A proposal to use axiomatic design and Product Platform Design to design a temporary shelter was formulated by Gilbert et al. [13]. It was appreciated that this example demonstrates the extent to which the combination of the two design methodologies can be used for an optimal solution to the problems raised by the realization of a complex civil engineering project. Puik et al. addressed the possibilities of combining the advantages offered by agile development and axiomatic design, respectively, aiming at harmonizing approximately contradictory design rules specific to the two methods of developing new products [47]. They appreciated that an attenuation of the agile design rules in the first phase and the axiomatic design rules in the subsequent phase would improve the design process.

Manufacturing tools and systems. The problem of using axiomatic design in the case of manufacturing tools and systems has been relatively often addressed by researchers. The first applications of axiomatic design aimed at improving manufacturing processes probably contributed to this situation. It can thus be seen that the principles of axiomatic design have been used to balance assembly lines [48], to develop total productive maintenance applied in manufacturing organizations [49], to develop manufacturing systems [50], etc.

Extensive and distinct approaches to manufacturing systems through the principles of axiomatic design have been described in several papers by Cochran et al. [51–53]. The problem of developing computer integrated manufacturing systems was addressed by Delaram and Valilai [54]. A point of view on axiomatic design in manufacturing systems was carried out by Rauch et al. in 2016 [5].

Energy. The heat ventilation and air-conditioning (HVAC) systems in commercial buildings could be improved, taking into account comfort and energy consumption using axiomatic design. Cavique and Conçalves-Coelho proposed practical solutions starting from the requirement corresponding to the first axiom, which implies the existence of independent or coupled systems [55].

Mechanical components. Probably a wide application of axiomatic design took place in mechanical structures and components. There are thus a wide variety of such structures or components in the design of axiomatic design principles.

Hydraulics and fluid mechanics. In principle, the components of hydraulic systems are also mechanical components, so they could be included in the topic addressed in the previous paragraph. Some specific problems of a hydrostatic spindle subsystem appeal to the use of the first axiom of AD [56]. The authors highlighted the importance of integrating the multisource information for the use of axiomatic design.

Electronics and electrical components. The conceptual design of mechatronic systems was the topic of research conducted by Chen and Jayram [57]. They developed an improved design methodology, starting from the principles of other methodologies, among them being axiomatic design.

The field of health. Optimization of patient flows in hospitals applying lean management principles, but applying a theoretical framework developed using axiomatic design was proposed by Arcidiacono et al. [58]. They considered that a group of patients with similar characteristics would contribute to better development of hospital activities.

Devices for older and disabled people. In a certain connection with the use of axiomatic design to solve health problems, it can be mentioned the identification of devices for elderly people or people with disabilities by applying axiomatic design principles. Thus, Mark et al. have invested efforts in designing worker assistance systems that can be used in the workplace by older people and respectively by workers with certain disabilities [59]. It is worth noting the use of one-on-one interviews to define the client's needs in this case clearly.

Agriculture and forestry. Sadeghi et al. found many work accidents in agricultural works deriving from the use of different solutions for power take-off of agricultural tractors [60]. They analyzed the existing alternatives of the power take-off subsystems and used axiomatic design to define functional requirements to reduce the risk of injury.

Jiang proposed a correlation of the axiomatic design process with the ontology information representation in the case of the development of small agricultural machinery products [61]. A reconfigurable product design system was considered.

Management in industry. Brown and Rauch analyzed the importance of functional requirements for promoting product creativity and sustainability when using axiomatic design [62]. The paper written by Brown and Rauch shows that "no design solution can be better than its FRs." They further considered that it is possible to select the most convenient design parameters by using axioms of axiomatic design. The integrated development of a product and its manufacturing process was the subject addressed by Vallhagen in his doctoral thesis [63].

The educational system. The systemic approach of the educational process, so the acceptance of the idea that there is an educational system, urged researchers to consider axiomatic design principles in the analysis and design of this system.

Thus, Mirzi and Liego-Betasolo evaluated the courses of materials engineering and fluid mechanics through an axiomatic design model [64]. In the case of his doctoral thesis, Towner has developed an interesting set of considerations according to which engineering education can be treated as a manufacturing system, and its problems can be solved efficiently using axiomatic design [65]. In a course at the University of Tokyo, Iino and Nakao used Design Record Graph and axiomatic design to identify and use students' creative resources [66].

Object handling and conveyors. Some problems regarding the handling and transport of products made from forest residues have been solved using the axiom of independence by Rodrigues et al. [67]. In this way, it became possible to equip better and organize the wooden pellets production line. Nadeo has proposed combining axiomatic design principles with a fuzzy logic approach to design an alternative propelled rear-wheel-drive vehicle chassis of car platform [68]. Khandekar and Chakraborty used fuzzy axiomatic design principles to select material handling equipment [69].

Services. The use of axiomatic design principles for developing knowledge management implementation services was proposed by Hao et al. [70]. They appreciated that in this way, a better collaboration of knowledge producers and receivers is possible and proposed the use of tools designed for this purpose. The possibility of using axiomatic design in financial services was noted by Banciu and Drăghici [71].

Mining and extraction. Zeng et al. considered the use of extended axiomatic design theory to the global mining supply chains, the latter appreciated as complex systems [72]. They considered that solving the problems specific to the global mining supply chains is difficult in the absence of methods capable of reducing the structural complexity of supply networks.

Illumination. The axiomatic design was used by Guls et al. to improve observation conditions around an autonomous underwater vehicle [73]. Based on the experience gained through previous research, a lighting module was proposed to be used to capture still images and video. An agile ergonomic monitor stand also involving a light source was proposed using the axiomatic design by Spalding et al. [74]

Breeding and fish farming. Vilbergsson mentioned axiomatic design as a solution to identify several possibilities to improve the specific functions of an intensive aquaculture system [75]. Using the axiomatic design theory, optimization of the solution of transfer bins for whole salmon grading has become possible [76].

Food and beverages. It is not surprising that axiomatic design principles are used in addressing food and beverage issues. Thus, an analysis of complexity in the kitchen was performed by Foley et al. [77], revealing the possibilities of using axiomatic design principles. Various issues specific to space life support systems, including food production, have been addressed through axiomatic design by Jones [78].

One can observe the large share of using axiomatic design to solve problems in the fields of industrial engineering and manufacturing engineering. A highlighting of the application of axiomatic design principles in different sectors of activity is possible, for example, by identifying papers published in these fields and indexed

	1991-1995				1996-2000					2001-2005					2006-2010					2011-2015					2016-2021						Total
Vehicles and vehicle's components,air/spacecraft				1		1		1		1	1	2	2	3	3	3	1	1	4	4	6	3	1	2	4	2	3	2		2	53
Chemistry and materials processing					1			1	1	2			2			3	2	2	1	2				7	2	1	1				28
Civil engineering										1							1					1		2	1	2	3	2		1	14
Manufacturing tools and systems				1	1		1	1		5	2	2	2	4	1	8	3	5	3	4	7	10	7	20	15	14	14	13	10	6	159
Energy														1			2	2	4	2	1		1	3		1	3		3	4	27
ICT and virtual environments				1									1	1		3	2	2	2	1			2	2	3	2	3			1	26
Algorithms and Software systems	1				1			1	2		1				1	2	2	5	3	2	2	2	2	5	4			1		1	38
Mechanical components				1								3	3	1	1	4		4	4	3	2		5	7	4	5	6		3	3	59
Hydraulics and Fluid mechanics									2	2	3	2	1	1		2	1	3	1	1	2		1	4	5	2	0	1	2		36
Electronics and electric components			1			1												1	1	2		3	3	1	2	1	2	1			19
Healthcare																			1			1	1	3	3	4	2	2			17
Education					1							1		1			1	2	2	2	1	1	1	2	4	2	3	1	1		26
Ergonomy															1		1		1		1	1		1		1		1			8
Agriculture and forestry														1	1										1						3
Management in industry		1		1					3					2	1		2	4		1	2	1	2	3	3	4	4	1	1		36
Robotics and Intelligent Manufacturing		1																2			1	1	1	2	10	2	3	1	4	1	29
Services (waste, transportation, watersupply)																			1		3	1	2	4	5	4	5	4	1	2	32
Other (sustainability, nuclear, economics, business)					1	1	2	2			3	1	3	2	2	7	7	12	6	7	4	6	5	10	16	9	17	7	11	6	147
Theoretical papers			1		1	3	1	2	3	2	5	5	6	11	7	10	18	13	14	17	7	7	7	14	29	6	17	13	5	2	226
TOTAL	1	2	2	5	6	6	4	8	11	13	15	16	20	28	18	42	43	58	48	48	40	38	41	92	111	62	86	50	41	29	984

Fig. 10 Use of axiomatic design in different fields, from papers published over time in Web of Science database

in the Web of Knowledge database. By considering some ideas from a previously mentioned paper [35] including other areas of activity in which axiomatic design was used, it was possible to develop the graphical representation in Fig. 10.

References

1. Suh NP (1990) The principles of design. Oxford University Press, New York
2. Suh NP, Bell AC, Gossard DC (1978) On an axiomatic approach to manufacturing and manufacturing systems. J Eng Ind Trans ASME 100:127–130
3. Suh NP (2001) Axiomatic design: advances and applications. Oxford University Press
4. Kulak O, Cebi S, Kahraman C. Applications of axiomatic design principles: a literature review. Expert Syst Appl 37(9):6705–6717
5. Rauch E, Matt DT, Dallasega P (2016) Application of axiomatic design in manufacturing system design: a literature review. Procedia CIRP 53:1–7
6. Sadeghi L, Houshmand M, Valilai OF (2017). Applications of axiomatic design theory in design for human safety in manufacturing systems: a literature review. MATEC Web of Conferences 127:01020
7. Heikkilä LJ (2020) Applications of axiomatic design in academic publications 2013–2018: a systematic literature review. School of Technology and Innovations Master's thesis in Industrial Management Master of Business
8. Brown CA (2020) Axiomatic design for products, processes, and systems. In: Matt D, Modrák V, Zsifkovits H (eds) Industry 4.0 for SMEs. Palgrave Macmillan, Cham

9. Thompson MK (2013) Improving the requirements process in axiomatic design theory. CIRP Ann 62(1):115–118
10. Thompson MK (2013) A classification of procedural errors in the definition of functional requirements in axiomatic design theory. In: Proceedings of the 7th International Conference on Axiomatic Design, Worcester
11. Brown CA (2011) Decomposition and prioritization in engineering design. In: Proceedings of the 6th International Conference on Axiomatic Design, Daejeon
12. Thompson MK (2014) Introduction to axiomatic design theory. In: Tutorials of the Eight International Conference on Axiomatic Design, Lisbon
13. Gilbert III LR, Farid AM, Omar M (2013) An axiomatic design based approach to civil engineering. In: Proceedings of the 2nd International Workshop on Design in Civil and Environmental Engineering, Worcester
14. Choi HJ (2005) A robust design for model and propagated uncertainty. Dissertation, Georgia Institute of Technology
15. Grozav I (2008) Improving quality through axiomatic design (in Romanian). Buletinul AGIR 1–2:105–111
16. Slătineanu L, Coteaţă M, Dodun O, Iosub A, Sîrbu V (2010) Some considerations regarding finishing by abrasive flap wheels. Int J Mater Form 3(2):123–134
17. Slătineanu L (2019) Fundamentals of scientific research (in Romanian). PIM Publishing House, Iași, România
18. Sundar PS, Chowdhury C, Kamarthi S (2021) Evaluation of human ear anatomy and functionality by axiomatic design. Biomimetics 6(2):31
19. Tomiyama T, Gu P, Jin Y, Lutters D, Kind C, Kimura F (2009) Design methodologies: industrial and educational applications. CIRP Ann Manuf Technol 58:543–565
20. Peace GS (1992) Taguchi methods—a hands-on approach. Addison-Wesley
21. Ranjit KR (1995) Primer on the Taguchi method. Society of Manufacturing Engineers
22. Teruo M, Shih-Chung T (2011) Taguchi methods: benefits, impacts, mathematics, statistics, and applications. ASME, New York
23. Filippone SF (1989) Using Taguchi methods to apply the axioms of design. Rob Comput Integr Manuf 6(2):133–142
24. Engelhardt F (2000) Improving systems by combining axiomatic design, quality control tools and designed experiments. Res Eng Des 12:204–219
25. Gohler SM, Frey DD, Howard TJ (2017) A model-based approach to associate complexity and robustness in engineering systems. Res Eng Des 28:223–234
26. Howard TJ, Eifler T, Pedersen SN, Gohler SM, Boorla SM, Christensen ME (2017) The variation management framework (VMF): a unifying graphical representation of robust design. Qual Eng 29(4):563–572
27. Kuo TC, Wang C-J (2019) Integrating robust design criteria and axiomatic design principles to support sustainable product development. Int J Precis Eng Manuf-Green Technol 6:549–557
28. Nepal B, Monplaisir L, Singh N (2006) A methodology for integrating design for quality in modular product design. J of Eng Des 17(5):387–109
29. Oh HL (2004) Unifying axiomatic design and robust design through the transfer function. In: Proceedings of the Third International Conference on Axiomatic Design Seoul
30. Rizzuti S, Gianipa F (2010) A mixed approach for robust design integrating Taguchi method in axiomatic design. In: Proceedings of IDMME—Virtual Concept, Bordeaux
31. Sarno E, Kumar V, Li W (2005) A hybrid methodology for enhancing reliability of large systems in conceptual design and its application to the design of a multiphase flow station. Res Eng Design 16:27–41
32. Yihai H, Zhao M, Wenbing C (2009) A technical framework of the taguchi system design method based on axiomatic design and TRIZ. In: Proceedings of the 2009 IEEE IEEM
33. Orloff MA (2010) ABC-TRIZ introduction to creative design thinking with modern TRIZ modeling. Springer
34. Gadd K (2011) TRIZ for engineers: enabling inventive problem solving. Wiley

35. Borgianni Y, Matt DT (2016) Applications of TRIZ and axiomatic design: a comparison to deduce best practices in industry. Procedia CIRP 39:91–96
36. Chechurin L, Borgianni Y (2016) Understanding TRIZ through the review of top cited publications. Comput Ind 82:119–134
37. Karampure R, Wang CY, Vashi Y (2021) UML sequence diagram to axiomatic design matrix conversion: a method for concept improvement for software in integrated systems. Procedia CIRP 100:457–462
38. Shirwaiker RA, Okudan GE (2008) Triz and axiomatic design: a review of case-studies and a proposed synergistic use. J Intell Manuf 19:33–47
39. Madara O (2011) Conceptual design using axiomatic design in a TRIZ framework. Procedia Eng 9:736–744
40. Yang K, Zhang H (2000) A comparison of Triz and axiomatic design. In: Proceedings of ICAD2000 First International Conference on Axiomatic Design Cambridge, MA
41. Fauzi MA, Humala NL, Rosnani G (2020) Comparison and integration of axiomatic design with quality function deployment as a design method: a literature review. IOP Conf Ser: Mater Sci Eng 1003
42. Yang J, Peng Q, Zhang J, Gu P (2018) Design of a hand rehabilitation device using integrated axiomatic and benchmarking methods. Procedia CIRP 78:295–300
43. Naddeo A (2004) Axiomatic design of a concept of car-platform for an electrical rear-wheel drive vehicle: a comparison with a fuzzy approach. In: Proceedings of the Third International Conference on Axiomatic Design 12
44. Tate D, Maxwell TT, Sharma BS, Patil K (2010) Selection of vehicle architecture for EcoCAR competition using axiomatic design principles. In: ASME International Design Engineering Technical Conferences and Computers and Information in Engineering Conference
45. Kazmer DO (2000) Axiomatic design of the injection molding process. In: Proceedings of the First International Conference on Axiomatic Design Cambridge, Worcester
46. Richards EV (2001) Design of an apparatus to measure gas solubilities in polymers. Master thesis, University of Toronto
47. Puik E, van Duijn J, Ceglarek D (2017) Guidelines for application of the constituent roadmap of product design based on axiomatic design. MATEC Web of Conferences 127:01013
48. Yilmaz ÖF, Demirel ÖF, Zaim S, Sevim Ş (2020) Assembly line balancing by using axiomatic design principles: An application from cooler manufacturing industry. Int J of Prod Manag Eng 8(1):31–43
49. Andemeskel F (2013) Total productive maintenance implementation procedures in manufacturing organizations using axiomatic design principles. In: Proceedings of ICAD 2013. The Seventh International Conference on Axiomatic Design Worcester, June 27–28, 2013 ICAD-2013-31
50. Aganovic D (2004) On manufacturing system development in the context of concurrent engineering. Doctoral thesis, Royal Institute of Technology, Stockholm
51. Cochran DS, Reynal VA (1996) Axiomatic design of manufacturing systems. The lean aircraft initiative. Report Series #RP96-05-14
52. Cochran DS, Eversheim W, Kubin G, Sesterhenn ML (2000) The application of axiomatic design and lean management principles in the scope of production system segmentation. Int J Prod Res 38(6):1377–1396
53. Cochran DS, Hendricks S, Barnes J, Bi Z (2016) Extension of manufacturing system design decomposition to implement manufacturing systems that are sustainable. J Manuf Sci Eng 138(10):101006
54. Delaram J, Valilai OF (2018) An architectural view to computer integrated manufacturing systems based on axiomatic design theory. Comput Ind 100:96–114
55. Cavique M, Gonçalves-Coelho AM (2009) Axiomatic design and HVAC systems: an efficient design decision-making criterion. Energy Build 41(2):146–153

56. Jia Q, Li B, Wei Y, Chen Y, Wang J, Yuan X (2016) Axiomatic design method for the hydrostatic spindle with multisource coupled information. Proc CIRP 53:252–260
57. Chen L, Jayaram M (2016) A preliminary study on conceptual design of mechatronic systems. Available at https://engineering.purdue.edu/~byao/Papers/AIM'01_Chen_UofT.pdf Accessed 15 Nov 2021
58. Arcidiacono G, Matt D, Rauch E (2017) Axiomatic design of a framework for the comprehensive optimization of patient flows in hospitals. J Healthcare Eng 3:1–9
59. Mark BG, Rauch E, Brown CA, Matt DT (2021) Design of an assembly workplace for aging workforce and worker with disabilities. IOP Conf Ser: Mater Sci Eng 1174:012013
60. Sadeghi L, Mathieu L, Tricot N, Al-Bassit L (2013) Toward design for safety part 1: functional reverse engineering driven by axiomatic design. In: Proceedings of the Seventh International Conference on Axiomatic Design
61. Jiang J, Xu F, Zhen X, Zhang X, Wang Y, Zhang L (2006) Axiomatic design using ontology modeling for interoperability in small agriculture machinery product development. In: Proceedings of PROLAMAT:184–191
62. Brown CA, Rauch E (2019) Axiomatic design for creativity, sustainability, and industry 4.0. MATEC Web of Conferences 301:00016
63. Vallhagen J (1996) An axiomatic approach to integrated product and process development. PhD thesis, Chalmers University of Technology
64. Llego-Betasolo M (2014) Axiomatic design model to assess influences affecting pedagogic-learning in the courses engineering materials and fluid mechanics. In: The Proceedings of the Eighth International Conference on Axiomatic Design, Lisbon
65. Towner W (2013) The design of engineering education as a manufacturing system. PhD Thesis, Worcester Polytechnic Institute (WPI)
66. Iino K, Nakao M (2016) Design record graph and axiomatic design for creative design education. In: The 10th International Conference on Axiomatic Design, ICAD 2016, Procedia CIRP, vol 53, pp 173–178
67. Rodrigues F, Fradinho J, Cavique M, Gabriel-Santos A, Mourão A (2019) An axiomatic approach to the design and operation of a wood pellet production line, ICAD 2019. MATEC Web of Conferences 301:00003
68. Naddeo A (2004) Axiomatic design of a concept of car platform for an electrical rear-wheel drive vehicle: a comparison with fuzzy approach. In: Proceedings of the Third International Conference on Axiomatic Design Seoul
69. Khandekar AV, Chakraborty S (2015) Selection of material handling equipment using fuzzy axiomatic design principles. Informatica 26(2):259–282
70. Hao Y, Kantola J, Valverde Arenas RR, Wu M (2013) Knowledge services in campus: the application of axiomatic design. In: Proceedings of the Seventh International Conference on Axiomatic Design, Worcester
71. Banciu F, Drăghici G (2003) About axiomatic design method. Acad J Manuf Eng 1(1):1–5
72. Zeng J, Zhu H, Kong J (2013) Enterprise architecture cybernetics for global mining projects: reducing the structural complexity of global mining supply networks via virtual brokerage. Adv Mat Res 634–638:3339–3345
73. Guls J, Bjarnason ÓI, Pétursson Ó, Einarsson SÖ, Foley JT (2016) Application of axiomatic design in designing autonomous underwater photography lighting. Procedia CIRP 53:278–283
74. Spalding C, Wei Z, Yarkov A (2019) Formulation of an agile office product: an application of axiomatic design in engineering. MATEC Web of Conferences 301:00008
75. Vilbergsson B (2016) Taxonomy and cross functions of technical solutions in aquaculture: resolving intensive aquaculture system treatment functions. MSc thesis, University of Iceland, Rewkjavik
76. Gerhard K (2016) Redesign of the SureTrack grader transfer bin using axiomatic design theory. MSc thesis, Reykjavík University

77. Foley JT, Puik L, Puik E, Smith J, Cochran DS (2019) Complexity in the kitchen. MATEC Web of Conferences 301:00007
78. Jones HV (2017) Axiomatic design of space life support systems. In: Proceedings of the 47th International Conference on Environmental System, Charleston, South Carolina

4D Printing on Textiles: Developing a File to Fabrication Framework for Self-Forming, Composite Wearables

Asterios Agkathidis and Guzden Varinlioglu

This design-led research investigates the development of self-forming wearable composite structures by 3D printing semi-elastic embossed patterns out of semi-elastic Thermoplastic Polyurethane 95 (TPU95) filament on pre-stretched textiles and releasing the stress after the printing has been completed. In particular, the study present and compare two methods of 'file to fabrication' techniques for generating self-forming textile shell structures: The first is based on printed patterns related to their stress line simulation and the second on modified geometrical patterns in relation to their curvature analysis. Furthermore, we will investigate the buckling degree of the composites in relation to their fabric thickness and elasticity. The findings emphasise the advantages and challenges of each method as well as presenting a comparative table chart highlighting the relationship between material properties, pattern geometry and thc formal vocabulary of the composite shells.

1 Introduction

At the beginning of the twentieth century, debates and developments which significantly changed the character of geometry gained momentum. By 1921, Albert Einstein questioned the relationship between geometry and experience, logical-formal language and real-life experience, and added physics as a complementary

---

A. Agkathidis (✉)
Liverpool School of Architecture, University of Liverpool, 25
Abercromby Square, Liverpool L69 7ZN, UK
e-mail: asterios.agkathidis@liverpool.ac.uk

G. Varinlioglu
Department of Architecture, Izmir University of Economics,
Sakarya Street, 156 35300 Balcova, Izmir, Turkey

P. Kyratsis et al. (eds.), *Advances in Product Design Engineering*,
Management and Industrial Engineering,
https://doi.org/10.1007/978-3-030-98124-2_3

thinking tool to understand the universe, which he called practical geometry [1]. The notion of geometry, which is meant to measure the Earth and initially connect it with realworld experience, turned away from reality into a logical and formal thought system, and in the early twentieth century, the relationship between geometry and experience became more questionable. Against the tendency to reduce geometry to logical and formal axioms, Buckminster Fuller proposed to reconsider orthogonal and cartesian geometry in relation to action, operation and movement [2]. He questioned geometry with different conceptions introducing the notion of folding as 'a way of thinking', 'folding as a form of action and operation' and 'folding by action'. Beyond the concept of geometry being reduced to static and rigid axioms, Buckminster Fuller placed movement at the centre, conceptualising it as self-provoking and self-initiating.

Frei Otto continued this approach with his research on doubly curved fabric structures during the 1970s and 1990s. Otto's design and form-finding process were strongly relying on physical models, rather than computational methods. Having developed a large range of innovative structures, by the early 1990s, Otto et al. [3] declared:

> Our times demand lighter, more energy-saving, more mobile and more adaptable, in short, more natural buildings, without disregarding the demand for safety and security.

In the early 2000s, others such as Brown and Rice [4] at Arup, focused on computational methods and techniques which they were applying for material innovation stress analysis and form-finding. Those developments continued in the 2010s with membranes and textiles being used successfully in building construction in the form of roofs, facades, pneumatic structures and tents. In our times, the rapid development of emerging technologies such as 3D printing and additive manufacturing, plus developments in material science, are enabling designers to consider further innovative solution synergies expanding its applicability to a wide facette of complexity and materiality such as plastics, concrete and metals. While 3D printing has been experiencing rapid development in the past two decades, the notion of four-dimensional printing has only appeared in 2012 according to Wu et al. [5]. The term describes the process through which a 3D printed object transforms its shape and structure over the influence of environmental parameters (e.g. temperature, humidity, light) or material properties (e.g. digital shape memory or stress relaxation), whereby the fourth dimension of the printing process becomes time.

In continuation to previous work by the authors on '3D printing of elastic fibre patterns on pre-stretched textiles' [6] and 'architectural hybrid material composites, computationally enabled techniques to control form generation' [7], this paper investigates the possibilities arising in shape, material properties and geometry of objects produced, by 3D printing of semi-elastic Thermoplastic Polyurethane 95 (TPU 95) onto pre-stretched elastic fabric (e.g. lycra-spandex) as shown in Fig. 1.

In particular, two different form prediction/form-finding methods were applied as described by Agkathidis et al. [6] and Berdos et al. [7] and tested in their effectiveness in predicting the desired shape and their suitability and limitations for producing particular geometries. Furthermore, the two methods were assessed and verified by

Fig. 1 This image illustrates the flat 2D pattern and resulting 3D geometry after release

using three different 3D printers (an Ultimaker-3 and the low-cost JG Aurora A5, Zaxe) allowing both to print directly on different fabrics, such as Polyamide Nylon and Elastane, Modal and Elastane composites, easily consumable wearables in the textile industry. Consequently, the following research questions were investigated:

- Which of the two proposed methods assessed here are more effective in controlling and predicting the form and performance of hybrid panels composed of semi-flexible, pattern fibres printed onto flat elastic, pre-stretched textiles?
- How do the material properties of the individual components—the textiles and the fibres—contribute to the properties of the composite material?
- How does the elasticity and thickness of the textile and the filament affect the degree of buckling of the final geometry?

To answer the above questions, a set of design-led, physical experiments were conducted using the two different form prediction methods on the three different printers and on different textile materials by developing a set of composite wearable prototypes. The findings were analysed and compared to enable conclusions.

2 Background and Literature

The study began by looking into the related work of other researchers in order to inform our research of the latest developments in the field. In their research, Joshi et al. [8] presented various active materials, 4D printing techniques and shape memories, however, their approach was mostly emphasising the field of structural engineering as they were focusing less on design. Cheng et al. [9, 10], describe the development of an additive manufacturing method combined with fused granular fabrication capable of producing 4D printed meta-structures, out of biocomposite material, which can change their geometry from flat to curved in relation to the environmental humidity. However, their work is using a completely different material pallet, to the ones examined in this paper.

Meyer et al. [11] investigated the adhesion of 3D printed polylactic acid (also known as PLA) on textile fabrics while Redondo et al. [12] researched the adhesion of 3D printed PLA samples on fabric by using the Fused Deposition Modeling technique. Even though both works provide valuable insights into the material properties and behaviour of PLA printed on fabric, they are neither examining the formal behaviour of the 3D printed objects nor their capability to change in time.

'Additive Manufacturing and Textiles' by Sitotaw et al. [13] broad overview casts light on various 3D printing techniques related to textiles, however, it was mostly focused on understanding material and technique properties rather than introducing novel materials and methods. Ehrmann and Ehrmann [14], presented their research on the 'Shape-Memory Properties of 3D Printed PLA Structures' which similarly to Giglio et al. [15] focuses on 3D printed PLA fabrics rather than on composites of PLA structures printed on textiles.

The prototypes produced in the workshop by Erioli and Naldoni [16] explored the possibilities in form generation by 3D printing semi-elastic PLA patterns on pre-stretched textiles. However, their investigation remained at an empirical level, which appears to emphasise artistic over empirical qualities, without incorporating simulation methods and form-prediction mechanisms. A similar technique was previously presented by Guberan and Clopath [17], in their 'Active Shoes' project, where a 3D printed geometry on a pre-stretched textile surface allowed the creation of a controlled and predictable shoe. However, there it might have not been their intention to provide evidence of simulation or form-prediction tools being used either.

In their article 'Printing on Fabric Meta-Material for Self-Shaping Architectural Models', Jourdan et al. [18] described a novel method of 3D printing PLA bars on pre-stretched textiles, including a star-based pattern system as well as a novel technique to simulate and predict the final shape of the models. However, they were not revealing the software and the tools used which are certainly different to the ones applied in this research. Their work offers a useful opportunity to compare the different methods in relation to their effectiveness.

In addition, Koch et al. [19], gave an overview of recent techniques for the generation of 4D textiles made by additive manufacturing on pre-stressed textiles offering a valid database for categorising, evaluating and assessing the techniques and methods of our research.

In their 'FabriClick' article Goudswaard et al. [20] showed a method for interweaving push buttons into fabrics by using 3D printing and digital embroidery. Even though they were achieving similar effects as described in our research, they don't seem to be using form prediction techniques such as stress line simulation and curvature analysis in their design process. A similar approach is described by Kycia [21], in the research on 3D printing on pre-stressed fabrics in order to create textile composites and explore their potential applications as building envelopes. Kycia has explored PLA, as well as polyolefin filaments on smaller as well as scale prototypes. Kycia showed rather simple, hyperbolic paraboloid geometries, without presenting any computational, form predicting methods. Finally, the research described by Aldinger et al. [22] in the 'Tailoring Self-Formation' paper has common ground with our work. However, finite element analysis appears to be their main tool of form prediction. In their material studies, carbon fibre rods were knitted into the fabric and helped to better control the self-formation geometry but they are not 3D printing filament on the textiles in order to produce the composite shapes.

The conclusion deriving from the literature review on similar research is that 4D printing on textiles is an up and coming research field that is currently being investigated by many research groups around the world. However, even though researchers have applied various methods and techniques of 4D printing and form prediction, our research appears to offer an original approach to the field as it is applying materials and methods not described by any of the researchers.

3 Development of Materials and Methods

As previously described, two different form-prediction methods were applied and tested for developing pattern geometries which we were then 3D printed on pre-stressed textiles. By releasing the newly composed prototypes, the objects should self-form into the desired wearable shape. Both methods have been developed using parametric tools (*Rhinoceros and Grasshopper*). In particular, Method 1 is based on utilising the Mean curvature analysis of the digital design model and adjusting geometric patterns on it by using an algorithm incorporating the Panelling Tools plug-in for Grasshopper (Fig. 1). The modified pattern is then being flattened, embossed and printed onto the pre-stressed textile.

Method 2 is based on an algorithm incorporating the *Karamba* structural simulation plug-in for *Grasshopper* [6], capable of conducting stress line simulations on the desired, digital design model (Fig. 2). The stress lines were rationalised and converted into a pattern which was then flattened, embossed and printed onto the pre-stressed textile.

Fig. 2 Diamond pattern adjusted according to Mean curvature map of a curved surface (Method 1)

Both methods were tested by conducting four experiments, where three types of wearable objects, bracelet/coffee cup holders, hats and facemasks/extensions were designed and fabricated. The first set of experiments (1, 2, 3, 4) measured the displacement between digital and physical models, thus the effectiveness of each method was verified, as well as identified the parameters which may influence the form prediction/generation. The study utilised an Ultimaker-3 3D printer, semi-elastic Thermoplastic Polyurethane 95 (TPU 95) filament and a Lycra Spandex 240 gms textile with a 40–20% stretch in the *X* and *Y* directions. In addition, the second set of experiments (5, 6, 7, 8) will examine the buckling capacity of the composite objects in relation to the use of different textile types such as Polyamide Nylon and Modal with different percentages of Elastane and thicknesses as well as to the embossed filament pattern (TPU95) by applying Method 2. The study tested the effectiveness of two low budget printers Zaxe and JG Aurora A5 printers, semielastic, which will be utilised for the second set of experiments, as well as the assembly method on the textile (either directly printed on the fabric or being laminated to it afterwards).

4 Verification Through Design Experiments

4.1 Experiment 1

Experiment 1 examined the design of a cylindrical bracelet/coffee cup holder with a single curvature geometry which was developed using Method 1. A hexagonal and a diamond-shaped pattern is applied to the bracelet model and adjusted to its Mean curvature analysis map as illustrated in Fig. 3. The pattern density was increased in the flattest areas (blue) and decreased in the areas with the highest curvature (red).

This experiment consisted of six variants, where pattern parameters such as rod thickness, pattern shape as well as the existence (or not) of a boundary rod were being tested and compared to the original, digital 3D model (Fig. 4). Variant v1.1 was designed using a hexagonal pattern with a rod thickness of 2.5 mm without a boundary frame and did not bend to the desired shape. Variant v1.2 has a rod thickness of 2 mm and a diamond-shaped pattern while variant v1.3 used exactly the same pattern as v1.2 but its rod thickness is 2 mm. While v1.3 over-performed by curving more than expected, v1.2 did not curve enough. Variants v1.4, v1.5 and v1.6 were all alterations of the diamond pattern with a boundary, differing only in their rod thickness (2, 2.5, 1.8 mm) with v1.6 proving to be the closest to the preferred shape. The conclusion deriving from experiment 1 was that Method 1 allowed the successful reproduction of the desired shape with variant v1.6 showing the smallest discrepancy to the digital 3D model (Fig. 5).

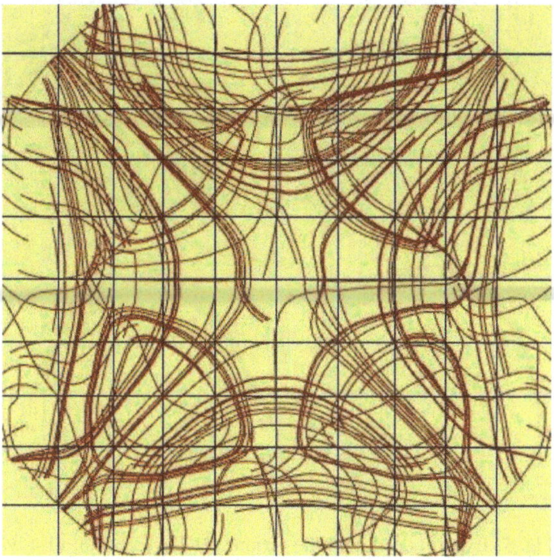

Fig. 3 Stress line simulation pattern produced on a curved surface (Method 2)

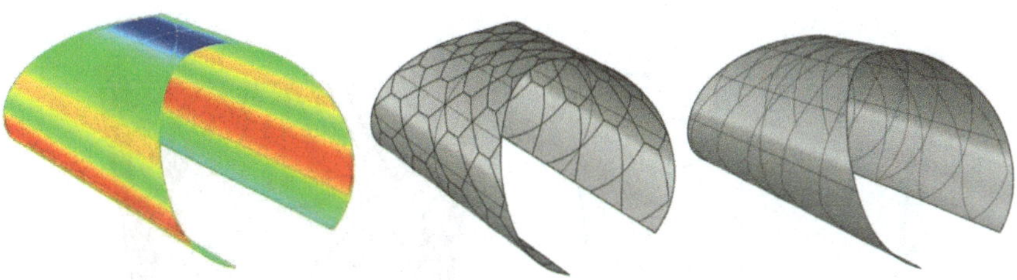

Fig. 4 Experiment 2, pattern adjustment on Mean curvature map (Method 1)

4.2 Experiment 2

Experiment 2 examined the design of a cylindrical bracelet/coffee cup holder with a single curvature geometry which was developed using Method 2. By applying the stress line simulation on the digital model, the stress line simulation pattern is generated as described in Fig. 6.

After being flattened, the stress lines were rationalised and transformed into a pattern that was embossed and 3D printed on the pre-stressed fabric. Two variants were tested. Variant v2.1 had a lower density stress line pattern with a rod thickness of 1.5 mm. Variant v2.2 had a higher density stress line pattern with a rod thickness of 1.5 mm. Both variations have an outer frame with a 1.5 mm thickness (Fig. 7). The conclusion deriving from experiment 2 was that Method 2 allowed the successful reproduction of the desired shape with variant v2.2 showing the smallest discrepancy to the digital 3D model (Fig. 7).

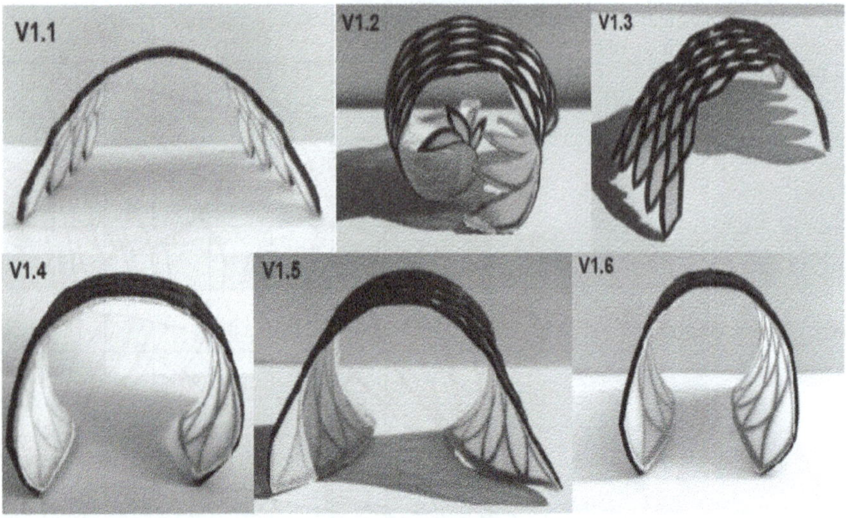

Fig. 5 Experiment 1, and the variants v1.1, v1.2, v1.3, v1.4, v1.5 and v1.6

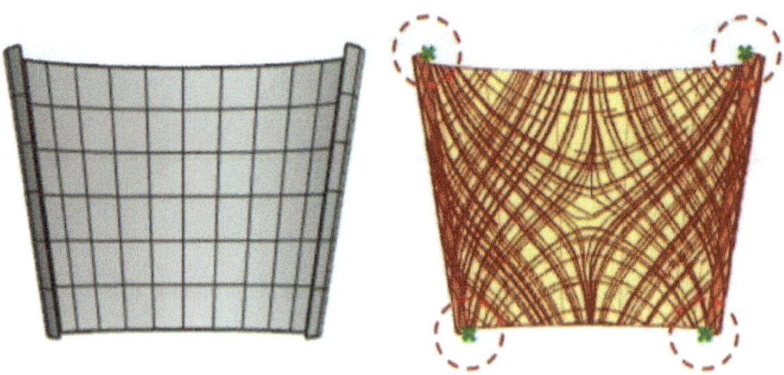

Fig. 6 Experiment 2, pattern generation via stress line simulation (Method 2)

4.3 *Experiment 3*

Experiment 3 examined the design of a facemask with a double curvature geometry which was developed using Method 1. We produced four variants as displayed in Fig. 8. Variant v3.1 is formed of a 3:3 diamond-shaped pattern with a rod thickness of 2 mm, variant v3.2 of a 4:4 diamond-shaped pattern with a rod thickness of 1 mm, variants v3.3 and v3.4 of a 5:5 and 6:6 diamond-shaped pattern accordingly, both having with a rod thickness of 1 mm. None of the variants has adopted a boundary frame.

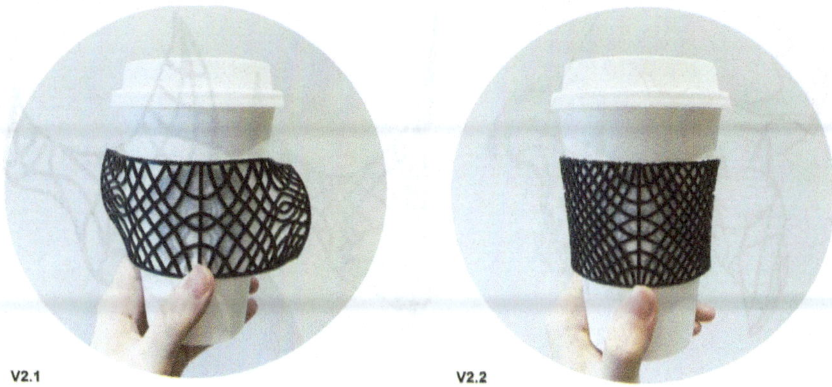

Fig. 7 Experiment 2, variants v2.1 and v2.2

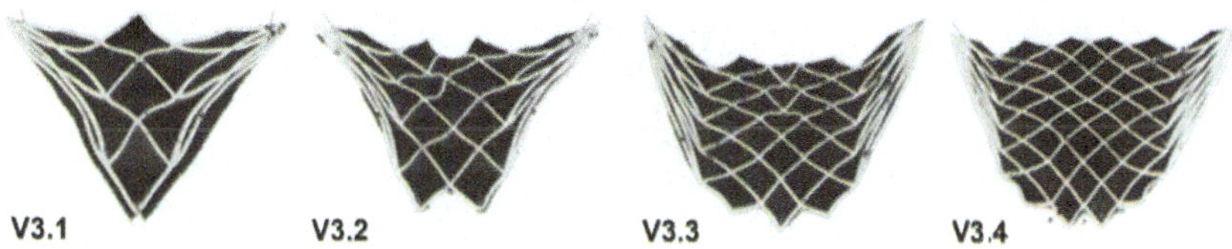

Fig. 8 Experiment 3, variants v3.1, v3.2, v3.3, v3.4

The conclusion deriving from experiment 3 was that Method 1 did not allow the successful reproduction of the desired shape, in particular in the *z*-direction, where the objects did not curve as expected.

4.4 Experiment 4

Experiment 4 examined the design of a facemask with a double curvature geometry which was developed using Method 2. Four variants were produced, as displayed in Fig. 9. Variant v4.1 applied a 40% stretch and has a rod thickness of 1 mm, variant v4.2 a 35% stretch with a rod thickness of 1.5 mm, variants v4.3 and v4.4 of a 30 and 25% stretch accordingly, both having a rod thickness of 2 mm. While variants v4.1, v4.3 and v4.4 have adopted a boundary frame, variant v4.2 has no boundary frame.

The conclusion deriving from experiment 4 was that Method 2 did allow the successful reproduction of the desired shape with variant v4.4 (including the boundary) showing the smallest discrepancy from the desired shape. Furthermore, variant v4.2 without a boundary border was the least successful variant in comparison to all other three variants which included a boundary border.

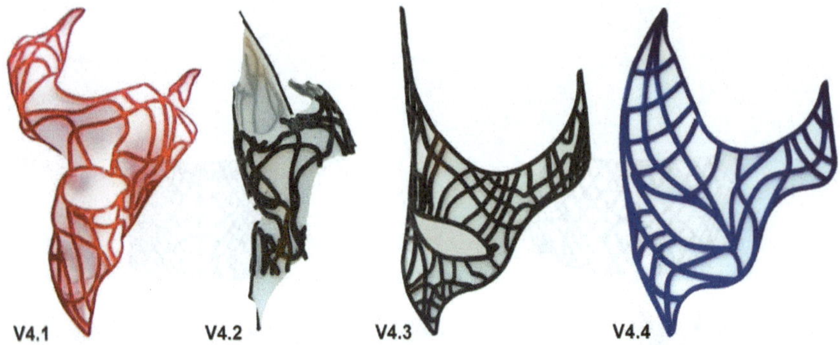

Fig. 9 Experiment 4, variants v4.1, v4.2, v4.3, v4.4

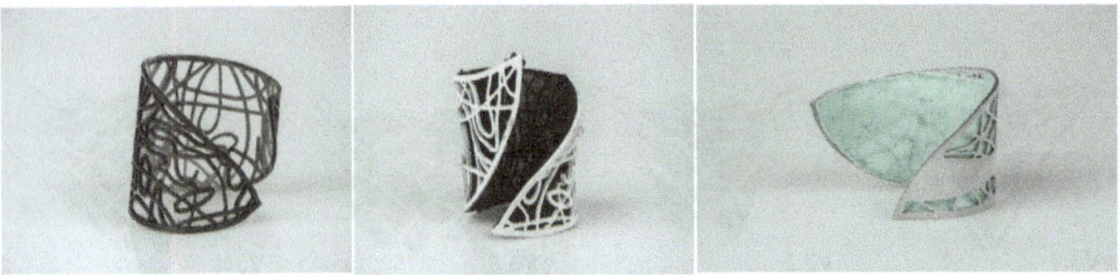

Fig. 10 Experiment 5, variants v5.1, v5.2, v5.3

4.5 *Experiment 5*

Experiment 5 examined the design of a cylindrical bracelet/coffee cup holder with a single curvature geometry using Method 2 and was printed on two different textiles with different stretching percentages. Variant v5.1 applied a 40% stretch of a silky matt transparent textile of 15 denier thickness (83% Polyamide, 17% Elastane), while v5.2 applied a 30% stretch of the matt opaque textile of 40 denier thickness (86% Polyamide, 14% Elastane) while v5.3 applied a 20% stretch of the same textile. All patterns have the same rod height (2 mm) but different rod thicknesses, v5.1 has a rod thickness of 1 mm, variant v5.2 with a rod thickness of 2 mm, variant v5.3 with a rod thickness of 3 mm (Fig. 10). The 3D prints of all variants v5.1, v5.2 and v5.3 were printed using a JC 3D printer directly on pre-stretched textiles. The conclusion deriving from experiment 5 was that all three textiles delivered a decent degree of buckling, which is strongly related to the rod thickness and density of the composite structure.

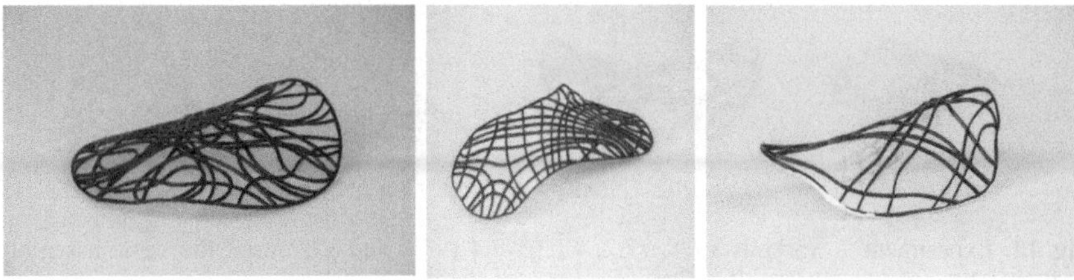

Fig. 11 Experiment 6, variants v6.1, v6.2, v6.3

4.6 *Experiment 6*

Experiment 6 examined the design of a face mask with a double curvature of a complete geometry (with boundaries) which was developed using Method 2 and will be printed on two different textiles using a variety of pattern geometries. All three variants (Fig. 11) were produced by the same printing method and the stretching factor applied was the same. Three similar textiles with a slight difference in thickness and the same rod thickness/height (1 mm of rod thickness and height) were used. Variant v6.1 utilised a sheer and shiny 80% Polyamide and 20% Elastane fabric with a thickness of 5 deniers. Variants v6.2 and v6.3 utilised an opaque mat of slightly thicker, 83% Polyamide, 17% Elastane textile with a thickness of 40 deniers. All three variants described closed, curvilinear shapes (circle, free form, and ellipse) with different 3D printed patterns. The distribution and density of the stress lines—equal in x and y dimensions—varied. The conclusion deriving from experiment 6 was that the thinner textiles delivered a higher degree of buckling, in comparison to the previous experiment (experiment 5).

4.7 *Experiment 7*

Experiment 7 examined a doubly curved face mask (with boundary borders) which was developed using Method 2 and was printed on two textiles with different thicknesses using different assembly methods. One of the two textiles used was 20 deniers thick composed out of 82% Polyamide and 18% Elastane, while the second one was 40 deniers thick composed out of 6% Polyamide and 14% Elastane. The embossed patterns had different thicknesses as well, but with the same rod height (Fig. 12). The textile of both v7.1 and v7.3 variants had a thickness of 20 deniers. Both v7.2 and v7.4 were applied on an opaque mat and slightly thicker textile (40 deniers). In both v7.1 and v7.2 variants, the patterns were printed directly on the fabric. Variants v7.2 and v7.4 were laminated on the fabric by using mitral Cyanoacrylate Adhesive glue. The distribution and density of the stress lines varied, as well as their rod thickness and height. The conclusion deriving from experiment 7 was that the objects where

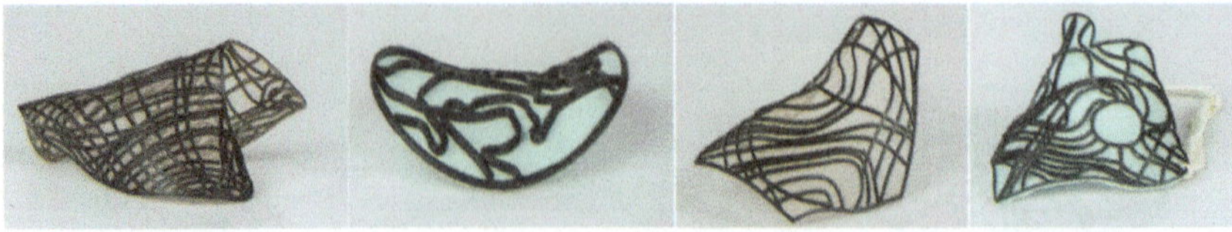

Fig. 12 Experiment 7, variants v7.1, v7.2, v7.3, v7.4 (v7.1 and v7.2 used the same assembly method, while v7.1 and 7.3 used the same textile)

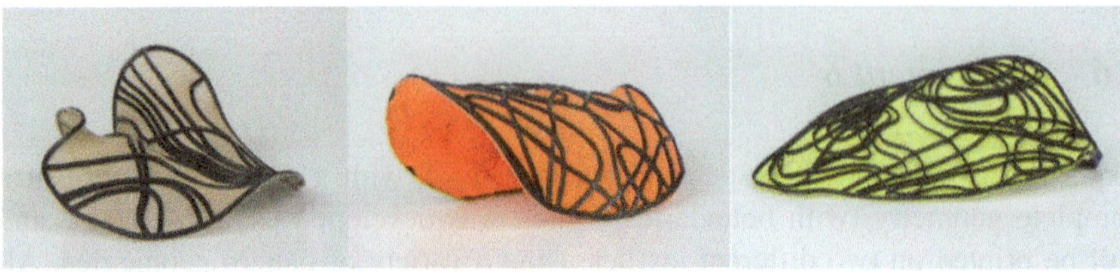

Fig. 13 Experiment 8, variants v8.1, v8.2, v8.3

the filament was printed directly on the textile have performed better (higher buckling degree) than those who were laminated on the fabrics after the printing had been completed (lower buckling degree).

4.8 Experiment 8

Experiment 8 examined a double curvature face mask that was developed using Method 2 and was printed on three different textiles with similar stretching percentages. Variant v8.1 was applied on a matt opaque tights fabric of 40 deniers thickness (92% Polyamide, 8% Elastane), while v8.2 was applied on a textile of 80 deniers thickness (91% Polyamide, 9% Elastane) and v8.3 on a modal textile (92% modal, 8% Elastane), all utilising patterns with the same rod height and thickness (Fig. 13). The 3D prints of all variants 8.1, v8.2 and v8.3 were printed using a Zaxe printer bed and the printed patterns were laminated on the pre-stretched fabric. Our conclusion from experiment 8 is that the thinner textiles delivered a higher degree of buckling.

5 Findings

The findings regarding the performance of the different methods are presented in Tables 1 and 2, a comparative displacement chart between variants and the digital 3D models used to design them. It became evident that variants v1.6, v2.2, v3.4 and

Table 1 Variation size displacement chart in mm

Variations	Stretch (%)	Material thickness (mm)	*X* dimension (mm)	*Y* dimension (mm)	*Z* dimension (mm)	*X*/*Y*/*Z* displacement (mm)
V1 DM	–	NA	30	85	40	–
V1.1	40	2.5	102	80	57	+72/−05/+17
V1.2	40	2.0	51	79	59	+21/−06/+19
V1.3	40	2.5	95	75	60	+65/−10/+20
V1.4	40	2.0	35	75	50	+05/−10/+10
V1.5	40	2.5	37	76	56	+07/−11/+16
V1.6	40	1.8	33	74	44	+03/−09/+04
V2 DM	–	NA	58	61	57	–
V2.1	40	1.5	124	61	60	+66/00/+03
V2.2	40	1.5	55	61	58	−03/00/+01
V3 DM	–	NA	170	95	106	–
V3.1	40	2.0	152	94	15	−18/−01/−91
V3.2	40	1.0	152	85	20	−18/−11/−86
V3.3	40	1.0	152	94	5	−18/−01/−101
V3.4	40	1.0	152	95	10	−18/00/−96
V4 DM	–	NA	134	150	43	–
V4.1	40	1.0	82	120	38	−52/−30/−5
V4.2	35	1.5	61	139	41	−73/−11/−2
V4.3	30	1.5	101	148	43	−33/−02/0
V4.4	25	2.0	124	148	43	−10/−02/−0

v4.4 were the most successful cases of each experiment. Both of the tested methods proved to be performing to an acceptable level.

Despite no prototype having the exact measurements to the digital model it has derived from, it appeared that Method 2, linked to the stress line simulation delivered the smallest discrepancies in the double curvature experiment (experiment 4), with variant v4.4 differing only −7% in the *X* direction and −1% in the *Y* direction while showing no difference in the *Z* direction (Table 2). The findings were very similar in the single curvature experiment (experiment 2), with variant v2.2 differing only −3% in the *X* and +2% in the *Z* directions while showing no difference in the *Y* direction (Table 2).

The findings regarding the second set of experiments (5, 6, 7, 8) and the buckling performance of different textile type related composites are demonstrated in Table 3. It became evident that thinner textiles with a high percentage of elastane provided a higher degree of buckling (e.g. variants v5.1 and v7.1 with a thickness of 15 and 20 deniers accordingly). Furthermore, it appeared that rod height and thickness have a significant impact on the buckling degree. Overall, we have observed that the right

Table 2 Variation size displacement chart in %

Variations	*X*/*Y*/*Z* displacement in %
V1.1	+240/−6/+40
V1.2	+66/−6/+47
V1.3	+216/−12/+50
V1.4	+16/−12/+25
V1.5	+23 −13/+40
V1.6	+10/−09/+10
V2.1	+113/0/+5
V2.2	−3/0/+2
V3.1	−10/−1/−85
V3.2	−10/−11/−100
V3.3	−10/−1/−95
V3.4	−10/0/−90
V4.1	−38/−20/−12
V4.2	−54/−7/−5
V4.3	−24/−1/0
V4.4	−7/−1/0

combination of textile thickness and composition, as well as the distribution of stress line pattern and rod dimensions, play an essential role in reproducing the desired form and have to be chosen accordingly (e.g. v6.2). Finally, in our experiments, it became evident that printing on pre-stretched textiles directly helps the variants to deliver a higher degree of buckling as laminated variants, such as v8.1 and v8.2 deliver a smaller degree of buckling.

6 Conclusions

Our conclusions will focus on answering our research questions. Which of the two proposed methods assessed here are more effective in controlling and predicting the form and performance of hybrid panels composed of semi-flexible, pattern fibres printed onto flat elastic, pre-stretched textiles? Even though both methods appear effective to a certain degree, Method 2 based on patterns generated by stress line simulation is more efficient, in particular, when the object is double curved. This becomes evident in Tables 1 and 2, where the most accurately reproduced objects are variants v2.2 and v4.4, both doubly curved. Method 1, based on the curvature analysis and penalisation, proved successful in reproducing single curved objects, e.g. variant v1.6. However, it failed to deliver enough buckling in the *z*-direction in experiment 2.

Table 3 Textile buckling degree chart

	Initial form	Curvature	Dimensions x/y	Buckling in the z-direction	Printer	Rod thickness/height	Textile thickness	Textile description	Fabric type	Method of assembly
v5.1	trimmed rectangle with curved borders	single curvature	20 cm × 8 cm	7 cm	JG	1 m × 2 mm	15 deniers	Silky matt opaque tights, flat seams and 3D technology for a perfect fit	83% Polyamide 17% Elastane	Printed on textile
v5.2	trimmed rectangle with curved borders	single curvature	20 cm × 8 cm	8 cm	JG	2 mm × 2 mm	40 deniers	Silky matt opaque tights, flat seams and 3D technology for a perfect fit	86% Polyamide 14% Elastane	Printed on textile
v5.3	trimmed rectangle with curved borders	single curvature	10 cm × 23 cm	10 cm	JG	3 mm × 2 mm	40 deniers	Silky matt opaque tights, flat seams and 3D technology for a perfect fit	86% Polyamide 14% Elastane	Printed on textile
v6.1	Full circle	double curvature	15 cm diameter	3.5 cm	JG	1 mm × 1 mm	5 deniers	Ultra-sheer and shiny look flat seams tights	80% Polyamide 20% Elastane	Printed on textile

(continued)

Table 3 (continued)

	Initial form	Curvature	Dimensions *x*/*y*	Buckling in the *z*-direction	Printer	Rod thickness/height	Textile thickness	Textile description	Fabric type	Method of assembly
v6.2	free form	double curvature	25 cm × 15 cm	2 cm	JG	1 mm x 1 mm	15 deniers	Silky matt opaque tights, flat seams and 3D technology for a perfect fit	83% Polyamide 17% Elastane	Printed on textile
v6.3	deformed ellipse with full borders	double curvature	16 cm diameter	5.5 cm	JG	1 mm × 1 mm	15 deniers	Silky matt opaque tights, flat seams and 3D technology for a perfect fit	%83 Polyamide %17 Elastane	Printed on textile
v7.1	Free form	double curvature	14 cm × 14 cm	10 cm	JG	2 mm × 1 mm	20 deniers	Sheer silky matt tights, flat seams	82% Polyamide 18% Elastane	Printed on textile
v7.2	Full circle	double curvature	15 cm × 15 cm	3.5 cm	JG	1.5 mm × 1.4 mm	40 deniers	Silky matt opaque tights, flat seams and 3D technology for a perfect fit	86% Polyamide 14% Elastane	Printed on textile
v7.3	Hexagon	double curvature	17 cm × 17 cm	11 cm	Zaxe	1.5 mm × 1.5 mm	20 deniers	Sheer silky matt tights, flat seams	82% Polyamide 18% Elastane	Laminated on textile
v7.4	Heart shape	double curvature	16.5 cm × 7.5 cm	8 cm	Zaxe	1.5 mm × 1.5 mm	40 deniers	Silky matt opaque tights, flat seams and 3D technology for a perfect fit	86% Polyamide 14% Elastane	Laminated on textile

(continued)

Table 3 (continued)

	Initial form	Curvature	Dimensions x/y	Buckling in the z-direction	Printer	Rod thickness/height	Textile thickness	Textile description	Fabric type	Method of assembly
v8.1	Ellipse	double curvature	17 cm diameter	2.8 cm	Zaxe	2 mm × 2 mm	40 deniers	Matt opaque tights	92% Polyamide 8% Elastane	Laminated on textile
v8.2	Full circle	double curvature	17 cm diameter	6.5 cm	Zaxe	2 mm × 2 mm	80deniers	matt opaque tights with 3D technology, water-resistant	91% Polyamide 9% Elastane	Laminated on textile
v8.3	Ellipse	double curvature	17 cm × 9 cm	3.3 cm	Zaxe	1 mm × 1 mm	Modal yarn	Transferring humidity very well	92% Modal 8% Elastane	Laminated on textile

How do the material properties of the individual components (the textiles and the fibres) contribute to the properties of the composite material? and how does the fibre pattern geometry influence the form of the composite hybrid panel? It appears that the relationship between textile type, rod thickness, stretching degree, pattern density and assembly method is very complex and particular, but also essential for reproducing the desired objects (Tables 1, 2, and 3). We could identify the following relationships: less dense patterns, operate best when directly printed on the textiles, on thinner and more elastic fabrics, while a bigger rod thickness and width is required in order to reproduce the object effectively.

How does the elasticity and thickness of the textile and the filament affect the degree of buckling of the final geometry? It appears that denser patterns, require thinner rod thickness and can deliver successful objects when printed on thinner, more elastic textiles. Overall, the elasticity and thickness of the textile are essential for both methods applied. The higher the elasticity degree of the textile, the more buckling is achievable (Table 3). The same applies to the textile thickness, where fabrics with a thickness of 5–20 deniers, perform much better than thicker fabrics (40 deniers and above). In contradiction to the formal vocabulary described by Aldinger et al. [22], the vocabulary described here is much more complex and polymorphous, as it varies in relation to the materiality of textile and fibres. In contradiction to their findings, we have observed that boundaries around the pattern geometries play an important role in the shape of the finalised panel.

This becomes evident in experiment 4, where variant v4.2 with no boundaries, did not manage to deliver the desired shape, while variants v4.3 and v4.4 which include a boundary came very close to the digital 3D model.

Furthermore, all three 3D printers (the Ultimaker 3 and the low budget printers JG and Zaxe) performed well and were able to 3D print on the textiles safely. Printing directly on the textile proved to be much more effective than laminating the pattern with glue, as the glued composites delivered less buckling in the z-direction (e.g. v8.1, v8.2 and v8.3, in Table 3).

Finally, one could highlight the variety of forms that are made possible by combining these two materials into a composite object; the semi-elastic Thermoplastic Polyurethane 95 and the elastic fabric. Forms that apply to rules and material properties, as well as to pattern geometry and design (Fig. 14). The success or failure of the final composite relies on the right proportion of design intentions and respect to the natural material memory and behaviour. This would allow us to enhance Frei Otto's call for lighter, more energy-saving, more mobile and more adaptable, in short, more natural buildings, or building components, such as roofs, ceilings, shading devices, tents, roofs and temporary shelters.

The limitations of this research project are linked to the size of all produced objects which is no bigger than 25 cm, which is the maximum printable size by the available 3D printers as well as a minimum rod thickness of 0.6 mm, linked to the minimum printable thickness by the printers. Our future plans include experimentation with larger-scale 3D printed objects, in order to verify our findings on a larger, architectural scale as well as examining the possibility of applying robotic technology for achieving more complex and reliable components.

Fig. 14 Variety of forms and design made possible by Methods 1 and 2

Acknowledgements We would like to express our acknowledgements to Darshan Chiba, Ffion Morris and Elliott Ward for executing experiment 1, Pattanan Inharwararak, Jinbi Jiang and Zili Qiu for executing experiment 2. Joseph Eyres, Viktor Lepetilo and Maxine Tai for executing experiment 3, Allena Anna Alphy, Diya Manoj Kulkarni and Poojapriyam Ravi from the Liverpool School of Architecture for executing experiment 4. Many thanks to Yorgos Berdos and Cesar Cheng for producing Fig. 1 and Mohannad Altabbal, Danial Mahabadi, Shreyans Mehta, Alejandro Ramirez, Ben Prescott, Jake Rothwell, Paul Nolan, Sami Samawi for producing the wearables on Fig. 14. Furthermore, our acknowledgements go to Furkan Sinan Üğütmen, Nezihe Gökçe Kökcan, Hilal Kaleli, Bilge Belenli, İrem Uysal, Yağmur Kaynar, Gözde Damla Turhan, Ece Küreli who acted as assistants for experiments 5, 6, 7 and 8 as well as Hugh Clarke, Filiz Keyder Özkan, Onur Dinmez, Çağlar Şendikici, Ayşe Bozkurt Karal from the Izmir University of Economics who executed experiments 5, 6, 7 and 8.

References

1. Einstein A (1921) Geometry and experience. Retrieved from https://www.relativitycalculator.com/pdfs/einstein_geometry_and_experience_1921.pdf
2. Massey J (2006) Buckminster Fuller's cybernetic pastoral: the United States Pavilion at Expo 67. J Archit 11(4):463–483. https://doi.org/10.1080/13602360601037883

3. Otto F, Rasch B, Schanz S (1995) Finding form: towards an architecture of the minimal. Edition Axel Menges, Berlin
4. Brown A, Rice P (2001) The engineer's contribution to contemporary architecture: Peter Rice. Thomas Telford, London
5. Wu JJ, Huang LM, Zhao Q, Xie T. (2018) 4D printing: history and recent progress. Chin J Polym Sci 36:563–575. https://doi.org/10.1007/s10118-0182089-8
6. Agkathidis A, Berdos Y, Brown A (2019) Active membranes: 3D printing of elastic fibre patterns on pre-stretched textiles. Int J Archit Comput 17(1):74–87. https://doi.org/10.1177/1478077118800890
7. Berdos Y, Agkathidis A, Brown A (2020) Architectural hybrid material composites: computationally enabled techniques to control form generation. Archit Sci Rev 63(2):154–164. https://doi.org/10.1080/00038628.2019.1666357
8. Joshi S, Rawat K, Karunakaran C, Rajamohan V, Mathew AT, Koziol K, Thakur VK, Balan ASS (2020) 4D printing of materials for the future: Opportunities and challenges. Appl Mater Today 18. https://doi.org/10.1016/j.apmt.2019.100490
9. Cheng T, Wood D, Wang X, Yuan PF, Menges A (2020) Programming material intelligence: an additive fabrication strategy for self-shaping biohybrid components. In: Vouloutsi V, Mura A, Tauber F, Speck T, Prescott TJ, Verschure PFMJ (eds) Biomimetic and biohybrid systems: living machines, vol 12413 (Lecture notes in computer science). Springer, Cham. https://doi.org/10.1007/978-3-030-64313-3_5
10. Cheng T, Wood D, Kiesewetter L, Oezdemir E, Antorveza K, Menges A (2021) Programming material compliance and actuation: hybrid additive fabrication of biocomposite structures for large-scale self-shaping. Bioinspi Biomim. http://iopscience.iop.org/article/10.1088/1748-3190/ac10af
11. Meyer P, Döpke C, Ehrmann A (2019) Improving adhesion of three-dimensional printed objects on textile fabrics by polymer coating. J Eng Fibers Fabr, January. https://doi.org/10.1177/1558925019895257
12. Redondo FL, Giaroli MC, Villar MA de AGO, Ciolino AE, Ninag MD (2020) Direct 3D printing of poly (lactic acid) on cotton fibers: characterization of materials and study of adhesion properties of the resulting composites. Macromol Symp 394:1900190. https://doi.org/10.1002/masy.201900190
13. Sitotaw DB, Ahrendt D, Kyosev Y, Kabish AK (2020) Additive manufacturing and textiles. Appl Sci 10(15):5033. https://doi.org/10.3390/app10155033
14. Ehrmann G, Ehrmann A (2020) Shape-memory properties of 3D printed PLA structures. In: Proceedings of the first international conference on "green" polymer materials, 5–25 November 2020, MDPI: Basel, Switzerland. https://doi.org/10.3390/CGPM2020-07198
15. Giglio A, Paoletti I, Conti GM (2021) Three-dimensional textiles in architecture and fashion design: a brief overview of the opportunities and limits in current practice. Appl Compos Mater. https://doi.org/10.1007/s10443-021-09932-9
16. Erioli A, Naldoni L (2017) Informed flexible matter workshop. http://www.co-de-it.com/WordPress/informed-flexible-matter.html
17. Guberan C, Clopath C (2016) Active shoes. https://selfassemblylab.mit.edu/active-shoes. MIT Self-Assembly Lab
18. Jourdan D, Skouras M, Vouga E, Bousseau A, (2021). Printing-on-fabric meta-material for self-shaping architectural models. In: Advances in architectural geometry, Paris, France, pp 1–19 ⟨hal-02925036⟩
19. Koch HC, Schmelzeisen D, Gries T (2021) 4D textiles made by additive manufacturing on pre-stressed textiles: an overview. Actuators 10:31. https://doi.org/10.3390/act10020031
20. Goudswaard M, Abraham A, Goveia da Rocha B, Andersen K, Liang R (2020) FabriClick: interweaving pushbuttons into fabrics using 3D printing and digital embroidery. In: Proceedings of the 2020 ACM designing interactive systems conference. Association for Computing Machinery, New York, NY, USA, 379–393. https://doi.org/10.1145/3357236.3395569

21. Kycia A (2019) Hybrid textile structures as means of material-informed design strategy. In: Ballestrem M, Borrego I, Fioretti D, Pasel R, Weidinger J (eds) CA2RE Berlin proceedings: conference for artistic and architectural (doctoral) research, Berlin. Universitätsverlag der TU Berlin, pp 34–35
22. Aldinger A, Margariti A, Koerner S, Suzuki A, Knippers, J (2018) Tailoring self-formation: fabrication and simulation of membrane-actuated stiffness gradient composites. In: Creativity in structural design. Boston. https://www.researchgate.net/publication/326552014_Tailoring_SelfFormation_fabrication_and_simulation_of_membraneactuated_stiffness_gradient_composites

Personalized Fashion On-Demand and e-Fashion Business Models: A User Survey in Greece

Evridiki Papachristou, Zoe Dimou, Margarita Grammatikopoulou, Lampros Mpaltadoros, and Thanos G. Stavropoulos

The pandemic is amplifying a trend that has been disrupting the clothing sector for decades: fashion on demand. Consumer needs and preferences are shifting globally to more personalized, on-demand fashion discovered and ordered online, which the COVID-19 mandated as the only means to shop for fashion, globally and for even up to a year. Fashion on demand reduces waste and reacts rapidly to trends. The objective of this study is to identify the reasons for which fashion on demand in Greece is only seen on a restricted scale, by investigating consumer attitudes towards e-commerce and their predispositions towards a new platform of personalized products. A questionnaire was constructed and answered by 550 participants. The outcomes confirm several of our hypotheses and present other interesting trends: consumers are willing to participate in the customization process and a collaborative e-business platform would be of benefit.

E. Papachristou (✉) · Z. Dimou
Creative Design and Clothing, School of Design Science, International Hellenic University, Thessaloniki, Greece
e-mail: evridikipapa@ihu.gr

M. Grammatikopoulou · L. Mpaltadoros · T. G. Stavropoulos
Centre for Research and Technology Hellas, Information Technologies Institute, Thessaloniki, Greece
e-mail: marggram@iti.gr

L. Mpaltadoros
e-mail: lamprosmpalt@iti.gr

T. G. Stavropoulos
e-mail: info@abotis.eu

T. G. Stavropoulos
Abotis Hellas PC, Thessaloniki, Greece

P. Kyratsis et al. (eds.), *Advances in Product Design Engineering*,
Management and Industrial Engineering,
https://doi.org/10.1007/978-3-030-98124-2_4

1 Introduction

The growth of demand from better educated consumers, mass customization, e-commerce, advances in integrated technology solutions and tools in the product development stage (e.g. ex-traction of body measurements, avatars, virtual try-on, AR/MR/xR applications) promote co-creation with the end-customer. The pandemic has affected the industry, and has created uncertainties that force clothing/textile companies to question and change their current practices [1]. An industry seen as a laggard in digitalization, will change dramatically in the next 15 years. A long-term goal is to bring production and consumption closer together both in space and time [2].

The contradictions between customer demand for personalized products and the relative short-age of personalized production have become increasingly prominent [3]. Indeed, though researchers have studied how the process of mass customization and on-demand business models could occur within the fashion's industry processes, the Greek fashion market and especially on-demand business models opportunities have not been previously addressed. This study aims to explore the potential of applying an on-demand business model in the apparel e-commerce environment by investigating consumers' interest in it. Furthermore, it is of interest to study if this way of operating can provide unique value to consumers in an efficient manner. Up to the authors' knowledge, a survey with 550 answered questionnaires on customer behaviour towards personalized clothing and co-creation has never been conducted in Greece before.

The COVID-19 pandemic amplified a trend that has been disrupting the clothing sector for decades: fashion on demand. Consumer needs and preferences are shifting globally to more personalized fashion, discovered and ordered online, as mandated by COVID-19 as the only means of shopping in the last year. In order to respond to trends and consumer demand, a new breed of start-ups for data analytics and adopt made-to-order production cycles is needed. Fashion on demand reduces waste and reacts rapidly to trends and consumer shifts in needs and behavioural change. Novel forms of personalization are enabled by Industry 4.0. According to Wang [3], direct customer input to design will enable companies to produce customized products with shorter cycle-times and lower costs than those associated with standardization and mass production.

2 Literature Review

This section presents related work in the field, from mass customization of textiles, on-demand manufacturing, to the current state-of-the-art in Greece.

2.1 *Mass Customization*

Fashion companies throughout the world have embraced mass customization in an attempt to overcome challenges like constant consumers' change of needs, global competitiveness, over-production [4], customized demand [5], ever-decreasing life-cycles, lack of information (customers' measurements), style preferences [6]. In order to deal with market uncertainty, mass customization has become an imperative measure for the industries as it gives flexibility and quick responsiveness to volatile market demand [7, 8]. Moreover, quick response supply is also helpful in reducing the environmental cost under the fashion mass customization system with consumer returns [9].

A major factor for the successful implementation of mass customization is customer involvement [10–13]. According to Chesbrough and Su [14, 15], intimate collaboration between end user and innovator is important, as customer knowledge is indispensable and leads to the development of innovative products. Gilmore [16] identified four distinct approaches to customization, namely collaborative, adaptive, cosmetic and transparent. The collaborative approach is based on a dialogue with individual customers, helping them articulate their needs, identifying the offering that fulfills those needs and making customized products.

Taking into account specific consumer needs and preferences is not new. Starkey targeted Baby Boomer Women [17]. Li developed a co-design system expected to apply in e-customization [18]. Kang found that a convenient co-design process, was an important attribute of favourable attitudes towards e-customized apparel even if the study concluded that perceived behavioural control was not a significant predictor of purchase intention of e-customized apparel [19].

2.2 *Technologies for Mass Personalization and On-Demand Manufacturing*

On-demand is about flipping the old mass model of producing many and hoping consumers will not only purchase it but actually fit into it; a consumer driven model which matches products to consumer demand. An ecosystem driven by consumers who are increasingly looking to purchase from non-mass-produced brands. Technology is the core of on-demand manufacturing and personalized clothing. Figure 1 demonstrates the processes involved in a make-on-demand business model, and the integrated technologies that enable the design, collaboration and launching of products in real time with the existing supply chain.

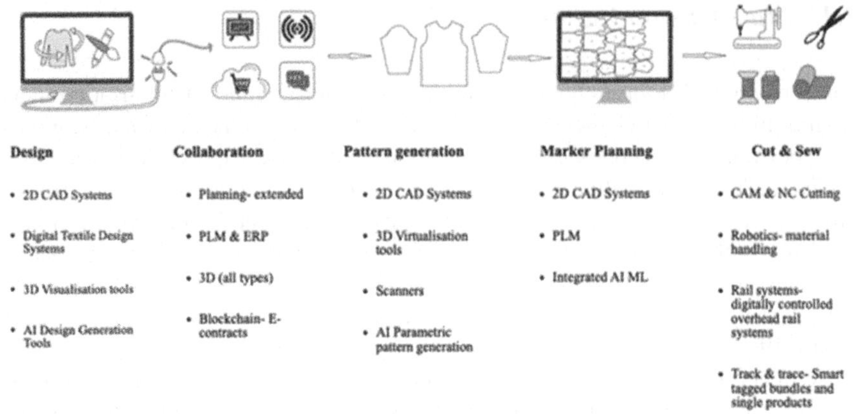

Fig. 1 Processes involved in a make-on-demand business model

2.3 *The Greek Textile and Apparel Industry*

Before examining the potential of bringing a new e-brand with on-demand characteristics, authors investigated other research studies on the Greek consumer regarding purchase of textile/apparel products. Kamenidou detected the underlying factors that affect the purchasing behaviour of Greek consumers without focusing on the consumers' reaction towards customization and new e-business models [20]. In the last decade, research studies on the Greek sector of textile and apparel are limited, restricted mostly as subjects of post-graduate thesis. Authors have searched in Google scholar & Scopus platforms, and selected the most relevant.

Koukovinos refers to psychosocial factors influencing young consumers' clothing disposal behaviour [21]. Perry focuses on the retail buyers in Greek luxury fashion brands [22]. Kaplanidou provides insight into the implications of the digital transformation in the apparel industry and Charamis focuses on the technology of ERP systems in the textile and apparel manufacturing companies [23, 24]. Xanthakou investigates consumers' relationships with Social Media, Tzavara explores the impact of Facebook and Instagram on consumer buying behaviour in the retail fashion market in Rhodes and Lalou showcases how data analytics can sup-port the 3PL decision making process on replenishing the network stores [25–27]. Papachristou identified how data analytics can be applied to the childrenswear market [28]. All these projects provide important data and help the reader understand the concepts of their research. However, they do not provide insights on fashion on-demand or the consumers' new behaviour characteristics.

OECD reported in 2019 that as digital transformation progresses, new business models arise in ways that are difficult to predict and that challenge traditional policy frameworks [29]. Contrastingly, Rainey, projects MarketWatch's expectation of the global on-demand manufacturing market to reach about $112 billion by 2024,

Table 1 Comparison between 2019 and 2020 on exports and imports of textiles and apparel (numbers in € millions)

	2020			2019		
	EU	Outside EU	Total	EU	Outside EU	Total
Exports						
Primary production	40	389	429	51	520	571
Textiles	242	132	385	301	120	421
Apparel	611	126	737	821	97	918
Total	903	644	1550	1173	736	1909
Imports						
Primary production	28	46	74	37	52	89
Textiles	247	293	540	268	316	524
Apparel	1011	715	1806	1284	1057	2341
Total	1287	1133	2420	1589	1425	3014

growing at 19.8% annually [30]. Fashion on-demand through e-commerce platforms is a new business model relying mostly on effective digital collaboration between involved parties, innovative business culture and on shifted consumers' shopping habits. The pandemic crisis has created new sales channels and new purchasing habits. Greece has shown significant limitations towards physical interaction and increase in e-commerce transactions. Although OECD reports that COVID-19 crisis is accelerating an expansion of e-commerce towards new firms, customers and types of products, data on consumer's behaviour towards a fashion on-demand e-commerce business model in Greece is missing [31]. The most recent data on the Greek clothing industry is provided by the Hellenic Apparel Association [32]. Table 1 shows the difference between the years of 2019 and 2020, regarding exports and imports of primary production, textiles and apparel. The supply chain of textiles and apparel recorded a decrease in turnover by 30%, retail sales by 23% and exports by 18.8%.

Greece is a market that needs further investigation and research on not only retail operations but more specifically in the possibilities of new e-business model adaptation that embrace fashion on-demand practices in order to take advantage of the increase of e-commerce transactions. Primary research data could help to face the COVID-19 crisis and contribute to a more sustainable industry that meets the demands from consumers for personalization and variety.

2.4 Current Situation

Several recent reports point out the phenomenon of individualism among new generations. Gartner finds that a growing number of brands are promoting items that appeal to consumers' demand for individual items (limited editions, collaborations)

[33]. McKinsey states that COVID-19 crisis resulted in years of online innovation and change that happened in a matter of months, as brands focused on generating revenue from the only channel available in many markets: e-commerce [34].

2.5 Personalization and On-Demand Case Studies

In 2017, Amazon patented an automated on-demand clothing factory [35]. Redcollar[1] focused on scaling customised transformation and developed the world's largest MTM (Made-to-Measure) advanced custom smart factory. The characteristic of the platform has built a new commercial civilization: everyone is designer, customer, operator and entrepreneur [3]. Buecher co-developed (with Adidas AG) a new production system closer to the customer (in-store) for knitted customized merino wool sweaters [36]. Twine Solutions'[2] new business model, produces only what the customer needs [37]. Popular brands in luxury fashion goods like Luis Vuitton and Coach provide online personalization services to their customers allowing them to select from a variety of colours, add signatures or monograms and special patches to the products [33]. Fast fashion market with examples of Calvin Klein and Reebok have also entered the customization initiative enabling customers to personalize underwear with text and patterns (Calvin Klein), or allowing them to vote on proposals drafted by Reebok's design team, and the brand will only manufacture a shoe if shoppers demonstrate demand [38]. Stitch Fix[3] and Net-a-Porter analyze browsing behaviour, interests and previous product selection and offer personalized outfit suggestions via data analysis technologies. Examples of on-demand manufacturing are Laws of Motion,[4] Stantt[5] and Redthread.[6] All three, solve the fit challenge with either a new sizing system or by made-to-order creation from mobile 3D body scanned data [39, 40].

Customer plays an important role in the on-demand business model; Gao, states that customers will be able to place an order and be involved in the entire process to produce and test the product as means of digital twin simulation technology [41]. According to Gribin, we are moving, however slowly and imperceptibly, from markets of millions to millions of markets of one: a market that will eventually be 'You' [39]. This is the concept that inspired the evolving of a similar idea for the market of Greece. Although the authors did a research on the available brands offering online custom clothing, discovered only two brands, which, in the time of writing this paper, are no longer working. However, taking into account the importance of on-demand and custom-made business models, the challenges that the Greek

[1] https://redcollar.ltd/.

[2] https://twine-s.com/.

[3] https://www.stitchfix.com/.

[4] https://lawsofmotion.com/.

[5] https://stantt.com/.

[6] https://redthreadcollection.com/.

textile and clothing industry face due to pandemic crisis and consumers' shift in purchasing habits, we continued the research, collecting primary data. In the next section, methods of doing so are presented.

3 Methods

3.1 Questionnaire Design and Development

The research presented was based on an online survey. The questionnaire, an outcome of research on similar surveys focusing on consumer behaviour, was developed and reviewed by the professor/project supervisor and a consulting group. Google forms was used to design the online questionnaire, manage its distribution and collection. Survey links were sent via email to students of the International Hellenic University. A total number of 600 questionnaires was distributed online with 550 returning completed and valid, leading to an excellent response rate (over 90%).

The final questionnaire consists of 17 questions. In the first section, socio-demographic questions are presented. The second section can be further divided into two parts. The first part, namely construct 'Attitude' (AT) contains six questions about the person's online shopping behaviour. These questions have been devised to measure consumers' attitude towards e-commerce. The second part, construct 'Satisfaction' (SA), comprises six questions and aims to examine the level of consumers' satisfaction concerning personalization in clothing products characteristics. The questionnaire is structured using Likert scale questions (from 1 = Not at all to 5 = To a large extent), permitting quantification of the participants' perceptions/views as they indicate the extent to which participants exhibit behaviours regarding e-commerce. The constructs of the questionnaire and their corresponding items are presented in Table 2.

3.2 Participants

The socio-demographic information of the survey respondents can be found in Table 3. The majority of the participants (72.5%) were female. The most participants (61.5%) belong to the age groups 21–30 years. Regarding their occupation, the participants comprise 32.2% students, 40.5% private sector employees followed by freelancers (16.7%), civil servants (7.8%) and un-employed participants (6.2%). More than 80% of the participants claimed an income between 0 and 999€. Monthly expenses on apparel are stated between 0 and 99€ in 74.2% of the answers.

Table 2 Constructs and corresponding items of the questionnaire

Construct (Latent variable)	Item [Factor loading]
Attitude (AT)	Q6: To what extent do you use the internet for purchasing apparel? [0.542]
	Q7: How often do you search for something specific while shopping online (e.g. I'm looking for a black jacket) [0.548]
	Q14: To what extent do you find it useful to be able to determine the design, colour and print of the garment you are interested in? [0.610]
	Q15: To what extent is delivery time important to you? [0.359]
	Q17: To what extent would you be willing to sacrifice speed in delivery time in order to obtain a product that meets your personal specifications/requirements? [0.702]
	Q18: To what extent would you be willing to spend more money in order to obtain a product that meets your personal specifications/requirements? [0.704]
Satisfaction (SA)	Q8: To what extent are you satisfied with the variety of colours, designs and prints you are looking for? [0.744]
	Q9: To what extent are you satisfied with finding your size in clothes? [0.663]
	Q10: To what extent are you satisfied with the final product in terms of fit regarding the size you have chosen? [0.748]
	Q11: To what extent are you satisfied with the combination of specifications you are looking for in a garment? (e.g. I am looking for a jacket with a … design, in … colour, in … fabric) [0.720]
	Q13: To what extent are you satisfied with the uniqueness an online store can offer? [0.655]
	Q16: To what extent are you satisfied with the delivery time when shopping online? [0.639]

3.3 Data Analysis

In order to assess the adequacy of the developed research instrument, statistical assumptions were tested. Also, the validity and construct reliability were evaluated [42]. Violation of the normality assumption occurs when the skewness and kurtosis values of each variable are greater than $+1.0$ or smaller than -1.0 [43]. Multi-collinearity exists when two or more independent variables correlate in a multiple regression equation. This statistical phenomenon is unwanted as it undermines the statistical significance of an independent variable. To avoid multi-collinearity, Field recommends that correlation coefficient between variables should not go beyond 0.8, a condition met in our dataset [44]. Additionally, the variance inflation factor (VIF) was used. A maximum VIF value of five was adopted according to the literature [45]. No multicollinearity issues were observed among constructs/variables. In Table 4,

Table 3 Socio-demographic information of the survey participants

Socio-demographic characteristics	Number of participants ($N = 550$)	Percentage (%)
Gender		
Male	151	27.5
Female	399	72.5
Age (years)		
15–20	50	9.1
21–25	183	33.3
26–30	155	28.2
31–35	72	13.1
36–40	41	7.5
Above 40	51	9.3
Occupation		
Student	177	32.2
Civil servant	43	7.8
Private sector employee	223	40.5
Freelancer	92	16.7
Unemployed	34	6.2
Monthly income (€)		
0–499	243	44.2
500–999	204	37.1
1000–1499	63	11.5
1500–1999	19	3.5
$\geq$ 2000	22	4
Monthly expenses on apparel (€)		
0–99	408	74.2
100–199	109	19.8
200–299	23	4.2
300–399	8	1.5
400–499	5	0.9
$\geq$ 500	1	0.2

the correlations and properties of all constructs and socio-demographic variables are presented.

There is no suspicion of common method variance. The questionnaire is short, the wording is un-complicated and concise, and examples are provided where needed. Anonymity is guaranteed and it is clearly stated that there are no correct/incorrect answers. The scale items were presented to the participants in a randomized order [46, 47].

Table 4 Correlations and properties of all constructs and demographic variables

	Attitude	Satisfaction	Gender	Age	Occupation	MI	MEA
Attitude	1	0.379**	0.023	−0.079	−0.004	0.052	0.216**
Satisfaction	–	1	−0.073	−0.021	0.049	0.106*	0.125**
Gender	–	–	1	−0.132**	−0.065	−0.308**	−0.100*
Age	–	–	–	1	0.525**	0.548**	0.046
Occupation	–	–	–	–	1	0.377**	0.106**
MI	–	–	–	–	–	1	0.305**
MEA	–	–	–	–	–	–	1
Mean	3	3	–	–	–	–	–
SD	0.6	0.6	–	–	–	–	–
Cronbach's alpha	0.666***	0.804***	–	–	–	–	–
AVE	0.35	0.49	–	–	–	–	–
Skewness	−0.166	−0.183	–	–	–	–	–
Kurtosis	0.427	0.910	–	–	–	–	–
VIF	1.054	–	1.125	1.577	1.266	1.669	1.235

Note MI, Monthly Income; MEA, Monthly Expenses on Apparel; *Correlation is significant at the 0.05 level (two-tailed), **Correlation is significant at the 0.01 level (two-tailed), ***Cronbach's alpha of the total scale: 0.787

The Kaiser–Meyer–Olkin measure of sampling adequacy (KMO = 0.822) and Bartlett's Test of Sphericity (<0.001) suggest that the sample is factorable. Factor analysis using varimax rotation method was conducted to study the sufficiency of the grouping categorization of the questions to form the desired constructs, AT and SA. Measurement items with factor loadings > 0.30 are considered to be significant, >0.40 more important and >0.50 very significant [48]. Table 2 shows the factor loadings for each item. The variance accounted for by the two factors is 45.3%. The visualization (using RStudio) shows that the groups reflect sufficiently the two categories defined, 'Attitude' and 'Satisfaction' (Fig. 2).

Furthermore, in order to understand whether the questions all reliably measure their corresponding latent variables, Cronbach's alpha was used. Cronbach's coefficient alpha of the total scale was found, 0.787, which can be considered satisfactory [48, 49]. For construct 'Attitude' it was found 0.666 and for latent variable 'Satisfaction' 0.804, indicating an acceptable level and a very good level of reliability respectively.

Next, the validity (convergent and discriminant) of the constructed scale was assessed. Average variance extracted (AVE) values should be higher or close to 0.5 for acceptable internal consistency (convergent validity) [50]. Discriminant validity is also confirmed, as the square root of the AVE exceeds the inter-correlations of the construct with the other constructs [51].

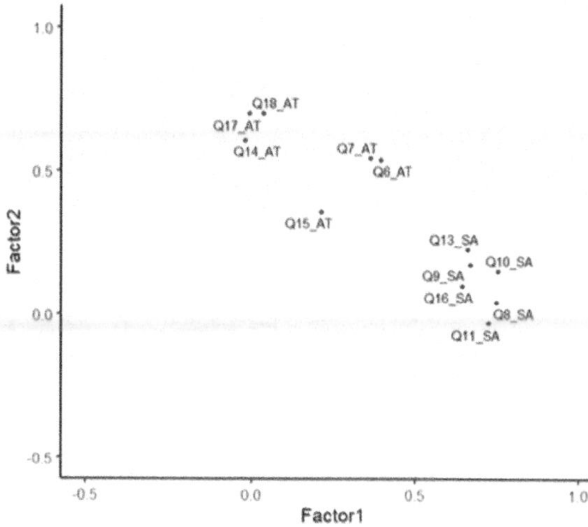

Fig. 2 Factor analysis, visualization of the results

Having demonstrated the adequacy of the formed constructs, the mean score of the multi-items was found for each construct. Hierarchical linear regression analysis was applied to explore if a relationship exists between the constructs AT and SA (Hypothesis A). Hypothesis A, investigates if the 'Attitude' of a consumer while shopping online affect his overall 'Satisfaction' towards e-commerce. In addition, we wanted to further investigate the existence of correlations between individual items of the questionnaire. For that, Hypotheses 1–8 were formed, where different questions were combined and correlations were tested. In this way, it would be possible to detect trends and better our understanding of consumers' perception towards e-commerce and customization of apparel.

For the analysis, the IBM SPSS 26.0 software was used. Descriptive statistics were used to analyze close-ended questions. As experimental data are generated from measurements taken from ordinal scales (categorical data), chi square test was employed to determine the significance of the association of different items/variables presented, and Spearman correlation analysis was employed to determine the strength of the relationship between selected variables.

4 Results

4.1 Testing Hypothesis A

The results of the hierarchical regression analysis are presented in Table 5. The formed Hypothesis A, was found statistically significant at $p < .05$ level. The satisfaction the consumers exhibited towards online shopping was significantly affected by their attitude regarding e-commerce ($\beta = 0.416$, $t = 10.165$). The demographic

Table 5 Results of testing Hypothesis A of the relationship between constructs AT and SA

Dependent variable	Independent variable	Std. coef. (β)	t-value	Sig. at $p <$ 0.05	Control variable	Std. coef. (β)	t-value	Sig. at $p <$ 0.05	Total $R2$	F-value (df1/df2)	Sig. at $p <$ 0.05
SA	Constant	–	9.455	<0.001	Age	−0.025	−0.500	0.617	0.175	103.331 (1/520)	<0.001
	AT	0.416	10.165	<0.001	Gender	−0.075	−1.772	0.077			
					Occupation	0.042	0.934	0.351			
					MI	0.043	0.840	0.401			
					MEA	−0.064	−1.435	0.152			

Note MI, Monthly Income; MEA, Monthly Expenses on Apparel

variables demonstrated no significant effects on consumer satisfaction. The explanatory power accounts for 17.5% variance ($R2 = 17.5\%$). As this study aims to investigate the existence of a relationship between latent behavioural variables, significant results with little explanatory power were considered tolerable [52, 53].

4.2 *Testing Further Hypotheses*

Prior to the survey a plethora of hypotheses was formed in order to assess the participants' attitude towards e-commerce in detail. The questions target to measure the ease of use, the possibility to customize the apparel and the importance of specific needs being met. The satisfaction the purchased apparel provides and the importance of delivery time were also assessed. Different correlations that could possibly occur between these questions and the demographic information of the participants were studied. The paired questions investigated for correlations and the corresponding hypotheses formed (H1-8) can be found in Table 6.

Table 6 also shows the results of the Chi square test and the Spearman correlation employed for the analysis. The values in bold display a statistically significant correlation showing which hypotheses can be supported while the others must be rejected.

For correct interpretation of these results, we further explored grouping the participant responses, per monthly income and per gender, and investigated their willingness to spend more money and their satisfaction with product fit in terms of size. Figures 3 and 4 show the distributions.

To further explore and visualize the correlation of responses between pairs of questions we demonstrate the scatterplots in Figs. 5, 6, and 7.

5 Discussion

It appears that the survey participants do not use the Internet as their primary source of purchasing apparel. Additionally, it is evident that the respondents are in favour of being involved in the process of clothing customization. Delivery time when shopping online is valued. Improvements concerning the customization of apparel would benefit the online shops and the clothing industry while at the same time they would increase the level of satisfaction reported by the consumers. The statistical analysis showed that five of the nine hypotheses can be supported (H1, H3, H5, H7, H8) while H4 is tending to yield statistically significant results. On the other hand, hypotheses H2, H2b and H6 could not be supported.

In more detail, the results showed that the higher the monthly income of the participant, the less important the cost increase of customized fashion clothes appears to be as they are willing to spend more money in order to meet personal specifications (H1). As the monthly income of the participants increases,

Table 6 Results of the Chi square test and the Spearman correlation

Hypothesis		Chi square test	Spearman correlation	
		*p**	*rs*	*p***
H1 (Q4 & 18)	The higher the monthly income, the less important is the cost increase of customized fashion clothes	**0.010**	0.143	**0.001**
H2a (Q2 & 14)	The younger the consumer, the more useful she/he finds it to participate in the customization of fashion clothing online	0.365	−0.055	0.198
H2b (Q3 & 14)	The occupation of the participant has an impact on how useful she/he finds it to participate in the customization of fashion clothing online	0.509	−0.024	0.587
H3 (Q1 & 10)	Female gender is more likely to report a low level of fit satisfaction	**0.041**	−0.082	**0.058**
H4 (Q8 & 14)	Low satisfaction in variety of colours, design & print, affects the response in the importance of consumer collaboration in the customization of apparel	**<0.001**	0.049	0.260
H5 (Q6 & 15)	The consumer who uses the internet for purchasing apparel, values higher the importance of time-to-deliver	**<0.001**	0.152	**<0.001**
H6 (Q5 & 14)	The lower the monthly expense for clothing, the less important to be part of apparel customization	0.121	0.035	0.419
H7 (Q6 & 7)	The extent to which the participant uses the Internet for shopping clothes relates to how often she/he searches for specific items of clothing while shopping online	**<0.001**	0.512	**<0.001**
H8 (Q17 &18)	The extent to which the participants are willing to sacrifice delivery time in order to obtain a product that meets their personal specifications, can be associated with the extent to which they are prepared to spend more money to acquire a product according to their personal specifications	**<0.001**	0.416	**<0.001**

*Asymptotic significance, 2-sided
**Significance 2-tailed

answers 'Not at all' and 'To a small extent' diminish. Specifically, 52% of the group with monthly income ≥ 2000€ answered with 'To a moderate extent', 'To a large ex-tent' while only 29% of participants with monthly income between 0 and 499 provided these answers (Fig. 3).

The female participants reported more frequently a low level of fit satisfaction in terms of size (H3). While men reported fit satisfaction 'To a moderate extent' in 29% of the cases, women replied in 18.9%. Also, fit satisfaction 'To some extent'

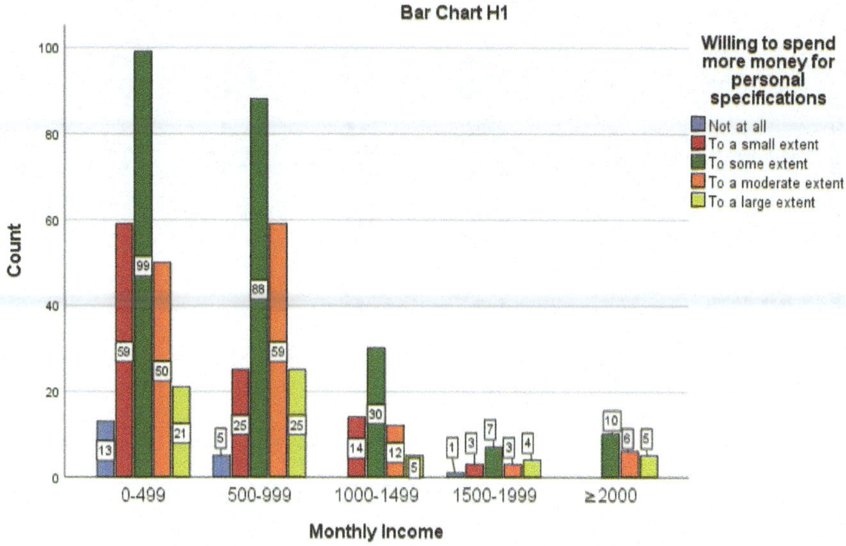

Fig. 3 Number of participants (Count) and willingness to spend more money to meet personal specifications, divided in five groups of monthly income (in €)

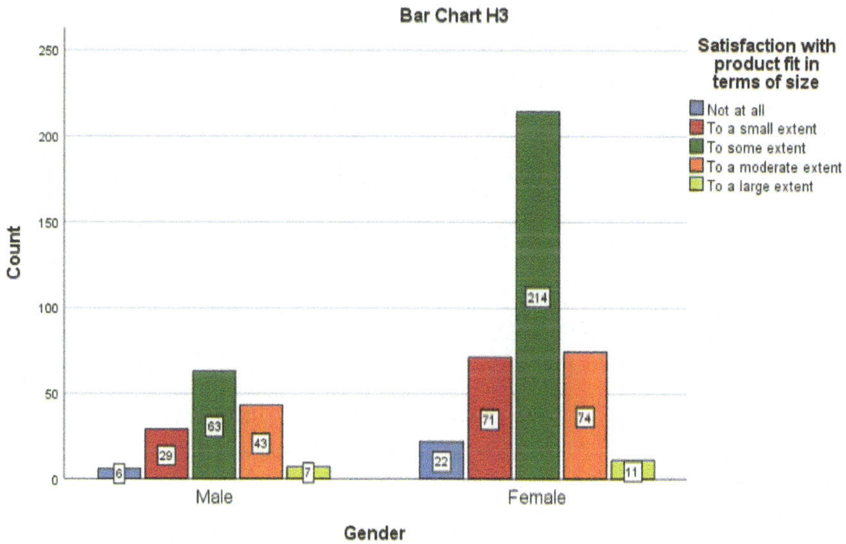

Fig. 4 Number of participants (Count) and satisfaction with product fit in terms of size, per gender group

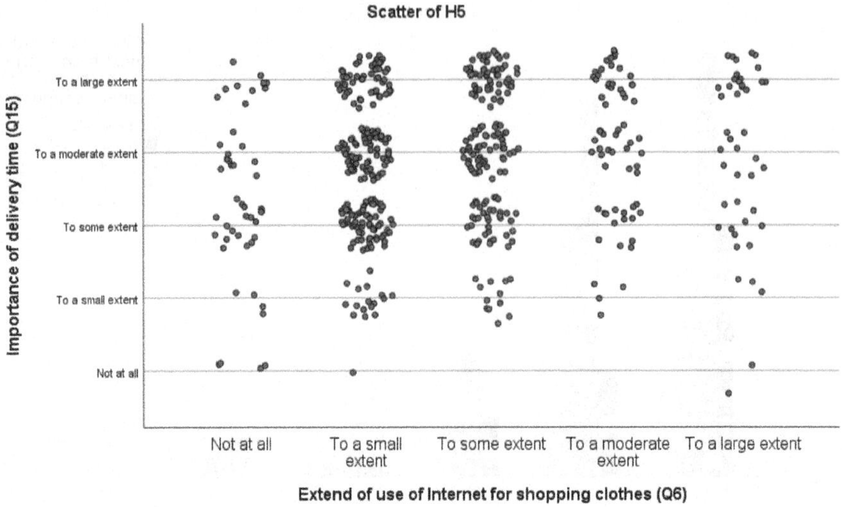

Fig. 5 Distribution of the number of responses per category across Q6 and Q15

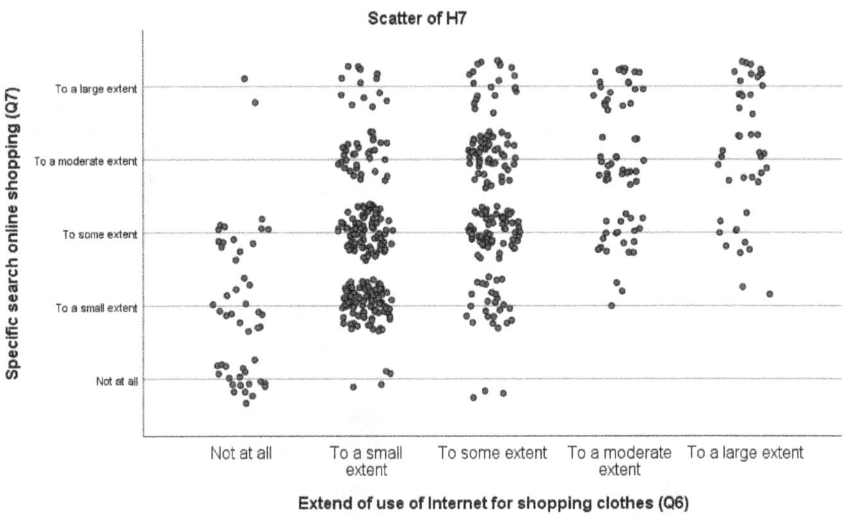

Fig. 6 Distribution of the number of responses per category across Q6 and Q7

was answered by women in 54.6% and by men 42.6%. Only 2.8% of the women participating responded 'To a large extent' with men appearing more content with 4.7%. The chart bar in Fig. 4 shows the participants' answers.

An interesting finding is the association between the extended use of the Internet for purchasing apparel and valuing the importance of the delivery time (H5). These two variables appear to have a positive correlation. The scatter plot in Fig. 5 presents

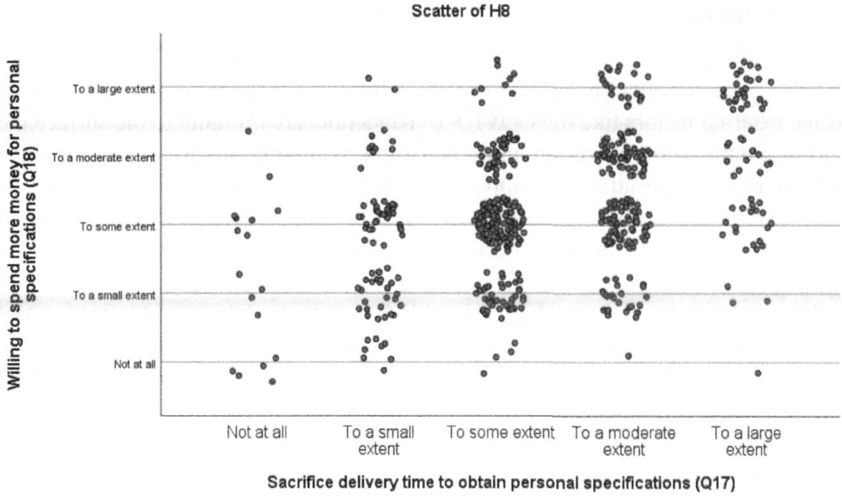

Fig. 7 Distribution of the number of responses per category across Q17 and Q18

the results. Excessive use of the Internet for shopping online affects the participants view on delivery time.

Also, the frequency of using the Internet for purchasing apparel correlates with searching online for specific items of clothing (H7). As it is visible from the scatter plot in Fig. 6, participants using the Internet for shopping apparel extensively, also claim that they are searching online for specific clothing items (e.g. black jacket). In detail, 40% of the participants that claimed they didn't shop clothing online answered 'Not at all' when asked to choose the extent to which they search specific items of clothing. On the other hand, 65 and 75% of the participants that use the Internet 'To a moderate extent' and 'To a large extent' respectively, also claim to search for specific items 'To a moderate extent' and 'To a large extent' while in those groups the answer 'Not at all' was not given.

Finally, the participants showed that they are willing to spend more money and sacrifice delivery time in order to purchase apparel that meet their personal specifications. These two variables show a positive correlation. The more willing participants are to sacrifice delivery time the more they are also willing to spend more money (H8) (Fig. 7). A near significant result was obtained from H4 concerning how low satisfaction in variety affects the response for importance of consumer collaboration in customization.

Age does not appear to have an impact on how useful the person finds it to participate in the customization of fashion clothing online (H2a). The data shows that in their majority, all age groups value the opportunity to determine different clothing features. The occupation of the participant does not have an impact on how useful he finds it to participate in the customization process (H2b). Furthermore, no significant correlation existed between monthly expenses and the importance to customize clothing (H6).

6 Conclusions

The change in consumer needs, the use of the Internet for shopping, as well as unexpected external factors like the COVID-19 pandemic, are bringing to the surface the field of clothing customization and co-creation. Although this trend is not new, it gains more and more attention. Apparel on-demand, can lead to important improvements regarding the clothing industry, such as reducing overproduction and resource exploitation. For this, a consumer driven business model supported by technology is needed.

A review of state-of-the-art highlights different aspects of mass customization and the current situation in Greece concerning this topic. A large scale survey ($N = 550$) of attitude and online shopping behaviour as well as requirements of Greek consumers shows interesting findings. There is a positive correlation between consumers' attitude (perception of using the internet for shopping clothes) and satisfaction towards purchasing apparel online. Furthermore, gender and monthly income were identified as influencing factors for fit satisfaction and willingness to spend a higher amount, respectively. Additionally, online shoppers can settle for longer delivery times and spend more money if the final product meets their requirements.

7 Limitations and Future Work

Through the survey findings, it was possible to gain a better understanding of consumers' attitude and their satisfaction towards online shopping. However, limitations were observed and should be taken into account in future work. First, regarding the questionnaire's audience. The conclusions drawn from the survey are limited to young people, students of the International Hellenic University (approximately 90% of the participants were under forty years old). It would be interesting to expand the survey participants' age range. The survey instrument and research model could benefit from revisions, construct 'Attitude' could be investigated further.

Additionally, we considered using Structural Equation Modelling (SEM) analysis to explore correlations between constructs. However, the questionnaire currently encapsulates only two main constructs: Attitude and Satisfaction. Further constructs could be added to enable us to investigate more relationships between them by using SEM.

References

1. Zhao L, Kim K (2021) Responding to the COVID-19 pandemic: practices and strategies of the global clothing and textile value chain. Cloth Text Res J 39(2):157–172
2. Walter L (2020) Digitalisation: EURATEX, The WTin Podcast. https://podcasts.apple.com/is/podcast/euratex/id1508352445?i=1000501415630

3. Wang Y, Ma H, Yang J, Wang K (2017) Industry 4.0: a way from mass customization to mass personalization production. Adv Manuf 5:311–320
4. Tseng M, Radke A (2011) Production planning and control for mass customization—a review of enabling technologies. In: Fogliatto F, da Silveira G, Mass customization (Springer series in advanced manufacturing). Springer, London
5. Yao J (2013) Scheduling optimisation of co-operator selection and task allocation in mass customisation supply chain based on collaborative benefits and risks. Int J Prod Res 51:2219–2239
6. Yeung H, Choi T, Chiu C (2010) Innovative Mass customization in the fashion industry. In: Cheng T, Choi T, Innovative quick response programs in logistics and supply chain management (International handbooks on information systems). Springer, Berlin
7. Silveira G, Borenstein D, Fogliatto F (2001) Mass customization: literature review and research directions. Prod Econ 72:1–13
8. Peters L, Saidin H (2000) IT and the mass customization of services: the challenge of implementation. Int J Inf Manage 20(2):103–119
9. Choi T, Guo S (2018) Responsive supply in fashion mass customisation systems with consumer returns. Int J Prod Res 56(10):3409–3422
10. Chandra C, Grabis J (2004) Managing logistics for mass customization: the new production frontier. In: Chandra C, Kamrani A, Mass customization: a supply chain approach. New York
11. Duray R (2002) Mass customization origins: mass or custom manufacturing? Int J Oper Prod Manag 22(3):314–328
12. Schwegmann V, Strube G, Willats P, Linck J, Boenigk A (2003) Flexible production and supply chain systems: generating value through effective customization. In: Proceedings of the 2nd interdisciplinary world congress on mass customization and personalization
13. Silva R, Rupasinghe T, Apeagyei P (2019) A collaborative apparel new product development process model using virtual reality and augmented reality technologies as enablers. Int J Fash Des Technol Educ 12:1–11
14. Chesbrough H (2003) Open innovation: the new imperative for creating and profiting from technology. Harvard Business School Press, Boston
15. Su C, Chen Y, Sha D (2006) Linking innovative product development with customer knowledge: a data-mining approach. Technovation 26(7):784–795
16. Gilmore J, Pine B (1997) The four faces of mass customisation. Harv Bus Rev 75(1):91–101
17. Starkey S, Parsons J (2019) Inclusive apparel design for baby boomer women. Fash Pract: J Desig Creat Process Fash Ind 11(1):81–104
18. Li P, Yu C, Wu C (2019) Customer-centered co-design modularization: the skirt design on mobile application. J Text Inst 110:1538–1544
19. Kang JYM, Kim E (2012) e-Mass customisation apparel shopping: effects of desire for unique consumer products and perceived risk on purchase intentions. Int J Fash Des Technol Educ 5(2):91–103
20. Kamenidou I, Mylonakis J, Nikoloyli K (2007) An exploratory study on the reasons for purchasing imported high fashion apparels: the case of Greece. J Fash Mark Manag 11(1):148–160
21. Koukovinos D (2012) Psychosocial factors influencing young consumers' clothing disposal behaviour in Greece. Master in Fashion Management with specialization in Fashion Marketing and Retailing. Boras, Sweden
22. Perry P, Kyriakaki M (2014) The decision-making process of luxury fashion retail buyers in Greece. J Fash Mark 18(1):85–106
23. Kaplanidou A (2018) Digitalization in the apparel manufacturing process. Master Thesis (U. U.-F. Geosciences, Ed.). Utrecht, The Netherlands
24. Charamis D (2018) Increasing competitiveness in the textile industry: a focus on the accounting benefits of erp systems by exploring cases from the UK & Greece. Theor Econ Lett 8:1044–1057
25. Xanthakou S, Antoniadis I (2019) Consumers' classification based on their behavior in the social media platforms of fashion companies. In: International conference on economic sciences and business administration, vol 5. Spiru Haret University, pp 228–235

26. Tzavara D, Clarke P, Misopoulos F (2019) An investigation of the impact of Facebook and Instagram on consumer buying behaviour: the case of retail fashion consumers in Rhodes, Greece. Int J Bus Econ Sci Appl Res 12(2):81–87
27. Lalou P, Ponis S, Efthymiou O (2020) Demand forecasting of retail distribution networks using data analytics and statistical programming. Manag Mark 15(2):186–202
28. Papachristou E, Bilalis N (2020) Data analytics and application challenges in the childrenswear market—a case study in Greece. In: Nyffenegger, F, Ríos, J, Rivest L, Bouras A, Product lifecycle management enabling smart X. PLM 2020, vol 594 (IFIP advances in information and communication technology). Cham, Springer
29. OECD (2019) Unpacking E-commerce: business models, trends and policies. OECD Publishing, Paris. https://doi.org/10.1787/23561431-en
30. Rainey B (2021) 3 benefits of on-demand manufacturing. https://www.mytotalretail.com/article/3-benefits-of-on-demand-manufacturing/
31. OECD (2020) OECD economic surveys Greece July 2020. http://www.oecd.org/economy/greece-economic-snapshot/
32. Greekfashion.gr (2020) Μείωση πωλήσεων 2020 στην ένδυση λόγω της πανδημίας. https://www.greekfashion.gr/en/news-trends/meiwsh-pwlhsewn-2020-sthn-endysh-logw-ths-pandhmias.278.html
33. Fiora L (2019) Gen Z gets personal (Gartner, Producer). Retrieved from Gartner: https://www.gartner.com/en/marketing/insights/daily-insights/gen-z-gets-personal
34. McKinsey Co & BoF (2020) The state of fashion 2021: in search of promise in perilous times. https://www.mckinsey.com/industries/retail/our-insights/state-of-fashion
35. Bain M (2017) Amazon has patented an automated on-demand clothing factory. Quartz. https://qz.com/963381/amazon-amzn-has-patented-an-automated-on-demand-clothing-factory/. Accessed 25 Apr 2021
36. Buecher D, Gloy YS, Schmenk B, Gries T (2018) Individual On-demand produced clothing: ultrafast fashion production system. In: Hankammer S, Nielsen K, Piller F, Schuh G, Wang N (eds) Customization 4.0. Springer proceedings in business and economics. Springer, Cham. https://doi.org/10.1007/978-3-319-77556-2_40
37. Moshe G (2019) On-demand and sustainable? https://www.whichplm.com/on-demand-and-sustainable/. Accessed 10 Jan 2020
38. Nishimuru K (2020) Reebok's 'first pitch' lets shoppers decide which sneakers get made. https://sourcingjournal.com/footwear/footwear-brands/reebok-first-pitch-sneaker-on-demand-manufacturing-personalization-footwear-224779/
39. Gribin E (2018) Is local-for-local or on-demand apparel production really happening? Sourc J. https://sourcingjournal.com/topics/thought-leadership/local-for-local-on-demand-apparel-production-115540/. Accessed 25 Apr 2021
40. Pardes A (2019) The perfect pair of pants is just a 3D body scan away. Wired. https://www.wired.com/story/bespoke-clothing-3d-body-scans/
41. Gao X, Xu J (2021) E-commerce in industry 4.0—chapter 10 in book. In: Intelligent information systems E-business in the 21st century, pp 235–267. https://www.worldscientific.com/doi/abs/10.1142/9789811231841_0010
42. Chi T, Kilduff PPD, Gargeya VB (2009) Alignment between business environment characteristics, competitive priorities, supply chain structures, and firm business performance. Int J Product Perform Manag 58(7):645–669
43. George D, Mallery M (2010) SPSS for windows step by step: a simple guide and reference, 17.0 update, 10th edn. Pearson, Boston, MA
44. Field A (2013) Discovering statistics using IBM SPSS statistics
45. Rogerson PA (2001) Statistical methods for geography. London
46. Chang SJ, van Witteloostuijn A, Eden L (2010) From the editors: common method variance in international business research. J Int Bus Stud 41:178–184. https://doi.org/10.1057/jibs.2009.88
47. Rodríguez-Ardura I, Meseguer-Artola A (2020) How to prevent, detect and control common method variance in electronic commerce research. J Theor Appl Electron Commer Res 15:1–5

48. Hair J, Black W, Babin B, Anderson R, Tatham R (2010) Multivariable data analysis, 7th edn. Pearson Education, New Jersey
49. Peterson RA (1994) A meta-analysis of Cronbach's coefficient alpha. J Consum Res 21(2):381–391. https://doi.org/10.1086/209405
50. Byrne BM (2016) Structural equation modeling with AMOS: basic concepts, applications, and programming. Routledge, London, UK
51. Fornell C, Larcker DF (1981) Structural equation models with unobservable variables and measurement error: algebra and statistics. J Mark Res. https://doi.org/10.2307/3150980
52. Cohen J (2013) Statistical power analysis for the behavioral sciences. Academic Press
53. Eisenhauer JG (2009) Explanatory power and statistical significance. Teach Stat: An Int J Teach 31(2):42–46

Machine Learning to Classify and Predict Design and Fabrication Solutions of Architectural Prototypes Driven by Sustainable Criteria

Odysseas Kontovourkis and Panayiotis N. Panayiotou

This chapter presents a methodology for using Machine Learning (ML) in an attempt to classify and predict design solutions of complex structures and products, considering static and geometric criteria as well as criteria of sustainable fabrication during the development of small-scale prototypes. In order to achieve this, a series of approaches are incorporated in the suggested methodology, including Topology Optimization (TO) analysis, design solutions development, post-processing analysis of the design solutions and ML investigation. The Artificial Neural Network (ANN) applied in this work achieves the classification and prediction of design solutions but also the prediction of new design solutions through a supervised neural network training and testing algorithm. The results derived from students' design solutions and then their classification and prediction of new ones, aspire to formulate a methodological framework, which can lead to the design and fabrication of structures and products that can be validated as best fitting solutions with reduced material usage and environmental impact.

1 Introduction

Nowadays, the introduction of integrated computational design and fabrication techniques has provided the opportunity for developing, exploring, and manufacturing complex structures and products at no extra production time and cost. At the same time, valuable solutions with superior quality and productivity rates have emerged,

O. Kontovourkis (✉) · P. N. Panayiotou
Department of Architecture, University of Cyprus, P.O. Box 20537, 1678 Nicosia, Cyprus
e-mail: kontovourkis.odysseas@ucy.ac.cy

P. N. Panayiotou
e-mail: panayiotou.n.panayiotis@ucy.ac.cy

P. Kyratsis et al. (eds.), *Advances in Product Design Engineering*,
Management and Industrial Engineering,
https://doi.org/10.1007/978-3-030-98124-2_5

achieving in parallel, reduction of their material volume and their environmental impact [1, 2]. Together with aesthetical and geometrical criteria involved, these techniques have provided the framework for developing viable outcomes in architecture but also other fields like product design engineering.

While a large number of techniques have been introduced and various projects have been already presented, little work has been done in regard to their evaluation. Most of the works dealing with the evaluation aspect of the design solutions have been based on well-known approaches such as Genetic Algorithms (GA) and Multi-Objective Genetic Algorithms (MOGA) optimization [3]. In these cases, the evaluation of solutions is based on structural and environmental objectives criteria and key performance indicators. Nevertheless, due to the nature of architectural design and fabrication processes, where subjective criteria are also part of the evaluation process, the incorporation of inputs and outputs that can lead to complex solutions is an equally important aspect, which until recently remains unresolved.

Currently, Machine Learning (ML) approaches [4], initially introduced in other fields like image recognition and prediction of consumers behaviour, have been also applied in architectural design and fabrication but only in a preliminary stage, which does not provide a comprehensive overview on their implementation. An important aspect in ML implementation is the use of large datasets and data-heavy workflows as the starting point of any evaluation process that also includes classification and prediction, factors that have a significant role in architectural design and construction due to the complex nature of the process involved.

In an attempt to contribute towards this direction, the following section discusses relevant works regarding ML in computational design and fabrication and provides a state-of-the-art review, which assists in the development of the presented research methodology.

2 Relevant works

2.1 Managing large datasets in computational design and fabrication

A large amount of data from a variety of sources that occur in the building sector leads to heavy models that need to be managed through non-conventional methods. New approaches and systems need to be followed by architects and engineers to effectively manage big data processes for the purpose of forming concise design information. In particular, the architecture sector could apply Machine Learning (ML) methods to further control data-heavy workflows [5].

Thomsen et al. [6] address the difference between typical methods and ML approaches for a design to fabrication workflow. In particular, the standard method in Architecture, Engineering, and Construction (AEC) practice involves the creation of digital models that are specific, created through a rule-based approach, or resolved through computation. In contrast to this, ML models are predictive and are trained through a range of exemplars. In this manner, the fabrication procedure and material frameworks could be encoded to handle high degrees of complication and resilience [7].

Rebala et al. [8] indicate that current conventional programming methods are ineffective for handling large amounts of labelled datasets. Specifically, the standard programming approach comprises two stages. The first stage is to build a detailed plan for the program, for instance, a set of stages or rules to solve a certain problem. The second stage is the execution of the design as a program within a computer language. This method could be difficult for a range of real-life problems that demand a large number of data points to be labelled within a dataset. In order to tackle this, ML is used for making a classification of these datapoints. Specifically, the algorithms of ML enable the system to learn the data that is labelled, for instance from a group of examples that indicate the performance of the program. ML algorithms are becoming more precise while the size of the dataset is growing larger. The aim of these algorithms is to learn patterns or a variation of rules from a classified dataset in order to achieve a precise prediction of the labels of data points that are not in the dataset.

2.2 Machine Learning (ML)

ML is a field of computer science that examines methods and algorithms for the automation of complex tasks that are difficult to be programmed using standard programming procedures. The main learning models of ML can be distinguished as supervised learning, unsupervised learning, semi-supervised learning, and reinforcement learning. In this section, a further explanation will be made to the most common terms which are supervised and unsupervised learning. The supervised learning model is provided with a variety of data points that are labelled with the correct answer to a problem. Then, the algorithm needs to learn the main features of each datapoint to distinguish the solution. In contrast to this, the unsupervised learning model is given with a set of data and the correct answer is not given. According to a large amount of data, the model could determine similarities. Furthermore, the algorithm will distinguish groups of similar features or the correspondence of a new feature with a current group [8].

Neural networks are the basis of several algorithms and applications of ML. Specifically, these are networks of interconnected artificial neurons, and their structure is influenced by the neuron network of a brain. Generally, a neural network is employed for creating supervised machine learning models for classification. In ML, classification refers to the detection of the category to which input is related with, between a collection of possible categories [8].

Deep Learning (DL) is a subcategory of ML, which at the same time is a sub-field of Artificial Intelligence (AI). AI is a generic term that applies to procedures that allow computers to replicate human actions. DL algorithms seek to make assumptions equal to humans, through a wise framework and a continuous analysis of data. In order to accomplish this, DL uses Neural Networks [9]. Deep Neural Networks contain a vast number of neurons with several layers. DL systems are efficient for learning unseen attributes and data structures due to the availability of a significant amount of data nowadays. Some of the current practices of DL are language translation, speech recognition, and image processing [8].

2.3 Machine Learning (ML) in architectural design

According to Tamke et al. [5], there are five emerging practices for ML in Architectural Design. These are, ML for the analysis of spatial design, ML for complex simulations, ML for dataset classification, ML for adaptive operation data, and ML for fabrication data. The following paragraphs are mainly focusing on ML for complex simulations, dataset classification, and fabrication data.

Peng et al. [10] created an ML framework that can recognize spatial forms. The practice that is used in this research is ML for dataset classification. Specifically, a network is trained from the classification of a range of predefined spatial compositions which are called seed-spaces. The classifier can then predict the form of a particular space, and through this procedure, a test could be made for a building to obtain statistics of its components within space. The classification workflow comprises of two stages, training, and service. The system initially conducts spatial sampling and therefore runs the result through the system, so that on one hand it trains the network and on the other hand it utilizes the pre-trained network to obtain an assumption. During the training stage, the data from the seed-spaces is used in order to train the network. Throughout the service stage, the network assumes the best identical seed-space for any input information tested from a particular space utilizing a similar methodology.

ML methods could also be applied in computational design and fabrication projects. One such project is Lace Wall which is a form-active hybrid system (FAHS) that comprises glass-fibre-reinforced plastic rods, textile cables, and custom-made high-density polyethylene (HDPE) components, to connect cables with rods. The practice related to this project is ML for complex simulations. Specifically, a supervised learning method with Artificial Neural Network (ANN) is used for pattern recognition in order to identify and study load distributions. ANN classification enabled complicated structures to be optimized in conditions that are difficult to be studied using current typical methods. In contrast to other classification methods, this type of network was capable to learn several uncommon parameters. This allowed flexibility and accuracy for the classification of once hidden data. In addition, it enabled possible reuse of the optimization results' database and the trained network for several computational processes of a design or numerous designs that use a similar structural system [5].

Nicholas et al. [11] created a workflow with a neural-network system, that can enhance conformal 3D printing and improve current robotic fabrication techniques. This project is relevant to the practice of ML for fabrication data. Specifically, the project examines conformal printing onto substrates of unknown geometry. The workflow comprises four steps. The first step is the registration of data using laser scanning equipment mounted on a robot that can capture the geometry of a substrate with complex terrain. Then, the point cloud data from the scanner is processed and encoded as an input to the neural network. In order to use three-dimensional information as training data, the geometric characteristics need to be interpreted into images that are possible to be analysed through a Neural Network [12, 13]. Moreover, the performance of the substrate is predicted by using the neural network. Finally, the prediction is used for activating the printing toolpath. A design experiment has been made that tests this workflow for the creation of façade panels. The material used for the test was Polyethylene Terephthalate Glycol (PETG) pallets. This material was extruded onto a wooden substrate of complex geometry and then formed the reinforcement pattern of the panels [11].

There is a range of existing ML platforms that help researchers to apply ML in several fields including Architectural Design. Tamke et al. [5] used TensorFlow along with Grasshopper software for a project called A Bridge Too Far. These tools were used during the learning process of an ANN for creating a predictive geometry before fabrication. Nicholas et al. [11] used a Conditional Generative Adversarial Network (cGAN) to predict the performance of surfaces with complex geometry. This prediction is activated within Grasshopper software while Python programming language and TensorFlow is running in the back end. Sanatani [14] created an ML framework that predicts the affective impact of enclosures in spatial design. The parameters of this framework were analysed using Rhino and Grasshopper with an ML plugin called Lunchbox.

The state-of-the-art review on ML in architectural design and fabrication shows its great potential for implementation, especially when the investigation involves large datasets and workflows and the evaluation of results are not solely based on common input and output results but involves contradicting information due to the complexity of the process involved. This work builds on existing knowledge on ML in order to formulate a research methodology with a twofold aim.

Firstly, through a number of design solutions developed in the context of digital design and fabrication courses in the undergraduate level, the study aspires to raise awareness and to provide technical skills in computational design and fabrication among the students of architecture.

Secondly, through the classification and prediction of results based on ML principles [15], the study aims to analyse solutions and to provide a range of sustainable outcomes that are distinguished in terms of material volume and fabrication time but also environmental impact potential.

3 Methodology

In order to achieve the abovementioned objectives, the design solutions refer to the development of a number of prototypes based on Topology Optimization (TO) results, using 6 different loading conditions but with the same geometrical boundaries of a structural beam as the subject of investigation. The results of loading and mass distribution in the body of structure in the form of 6 case studies, are given to the students as a starting point for parametric design investigation and physical production. Apart from the structural performance provided based on the 6 case studies, the development of design solutions considers fabrication criteria, attempting to minimize material and hence improve environmental impact of physical prototypes. The development of solutions is achieved either through a bottom-up or a top-down design logic, where each student designs a structural unit and then a system of units, producing overall and customized beam prototypes that respond to the TO results according to the case study under investigation. Then, students are asked to proceed to the physical production of their design solutions in 1:10 scale using a digital fabrication process that involves 3D printing of formworks and then casting using concrete as the material of implementation.

The results derived from this investigation are post-processed in order to extract output data that correspond to input information used during the development of design solutions. Input and output data of 30 design solutions are used in a supervised neural network algorithm, which classify and predict design solutions according to the complex beam structures developed based on specific loading condition inputs.

The prototypes, which have been developed by the students as part of the course, and the introduction of simple input and output data for training, classification and prediction purposes using a neural network algorithm, aim to analyse the complex models, which can be validated as best fitting results in regard to their minimum material usage and their lower environmental impact performance.

3.1 Topology Optimization analysis

The initial step of the process refers to the production of 6 case studies based on Topology Optimization (TO) analysis using the TopOpt plug-in [16] for Grasshopper [17] (plug-in for Rhino software [18]). The specific plug-in is based on a basic Topology Optimization procedure following the work by Ole Sigmund [19]. The goal is to develop a series of analysis results regarding optimized material distribution in a 2D domain, which are used by students to develop their physical prototypes in the next stage. The TO analysis is conducted based on the same geometrical boundary conditions as the starting point but with different input loading conditions that are differentiated in each case study. The analysis starts by specifying a rectangular boundary that represents a structural beam with dimensions $0.8 \times 5.00\ \text{m}^2$. This geometry is developed in Rhino software environment and set to Grasshopper for further analysis steps. The geometry is defined as the design space for domain discretization. Then, a series of parameters are specified including the loading conditions (force point), the support conditions (support points), and the optimization settings. In particular, the force point input component consists of the point's coordinates that the force is applied and the force direction, in this case vertical with negative direction according to Y axis. The support point input component consists of the points where the supports are applied. In this case, two support points are applied with a specific degree of freedom according to the X and Y axes. In the first support point, there is no freedom in X and Y directions. In the second support point, there is freedom in X direction with Y axis to be locked. The final input component refers to optimization settings and more specifically to volume fraction, penalization power and filter size, which are used as input settings in 2D Topology Optimization component. Table 1 demonstrates the six different case studies, representing various input data for analysis.

The 2D Optimization generates 6 different output results that are related to the material distribution and geometry. The latter undergoes further elaboration to represent the results of optimization in the form of an outline geometry. This is achieved through the component Contour 1Material with 0.7 the input value of the density filter. Figure 1 demonstrates material optimization results of case study 6 in the form of an image representation and Fig. 2 represents the 2D representation of TO in the form of contours.

3.2 Design Solutions (DS) development

The optimized material distribution in 2D domain of the 6 case studies was given to 30 undergraduate students of architecture as the starting point for their design and physical production processing based on the use of parametric design principles. In order for the students to understand the purpose and objectives of this exercise but also to precisely follow the suggested design procedure, the design workflow was explained in a number of steps based on a top-down and bottom-up design logic. The exercise aimed to develop design solutions (DS) that consider the results of TO, and more specifically referred to the placement of a larger amount of material in the areas indicated by the TO analysis, minimizing at the same time the amount of material in other areas with further distance from the contours. The concept has followed principles found in functionally graded structures, where graded lattice structures as infill patterns were generated based on Topology Optimization analysis, aiming at additive manufacturing of their designs [20–22].

Based on a similar concept, instead of infill patterns, the students were asked to develop individual structural units for each one of the 100 cells enclosed in the 2D domain. These structural units were distributed in the 2D domain area of each case study under investigation and their parameters were differentiated according to their distance from the contours (see Fig. 2). Briefly, the two parametric design logics were described in the following steps.

Table 1 Input data for the 6 case studies developed using Topology Optimization

Case study	Design space (m^2)	Force point		Support points				Optimization settings	
		Position (0 to 1)	Direction (V)	Left point (Dof X, Dof Y)		Right point (Dof X, Dof Y)		Volume fraction (VolFrac)	Filter size (Rmin)
1	5.00 × 0.25	0.5	−1	False	False	True	False	0.4	2
2	5.00 × 0.25	0.25	−1	False	False	True	False	0.4	2
3	5.00 × 0.25	0.25	−1	False	False	True	False	0.6	1.5
4	5.00 × 0.25	0.5	−1	False	False	True	False	0.6	1.5
5	5.00 × 0.25	0.62	−1	False	False	True	False	0.4	2
6	5.00 × 0.25	0.62	−1	False	False	True	False	0.6	1.5

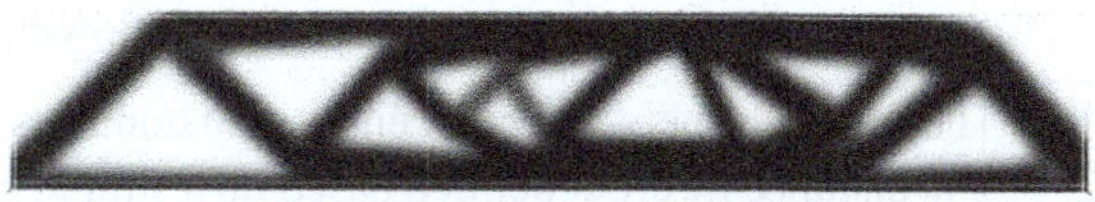

Fig. 1 Material optimization based on TO extracted from the TopOpt plug-in for the case study 6

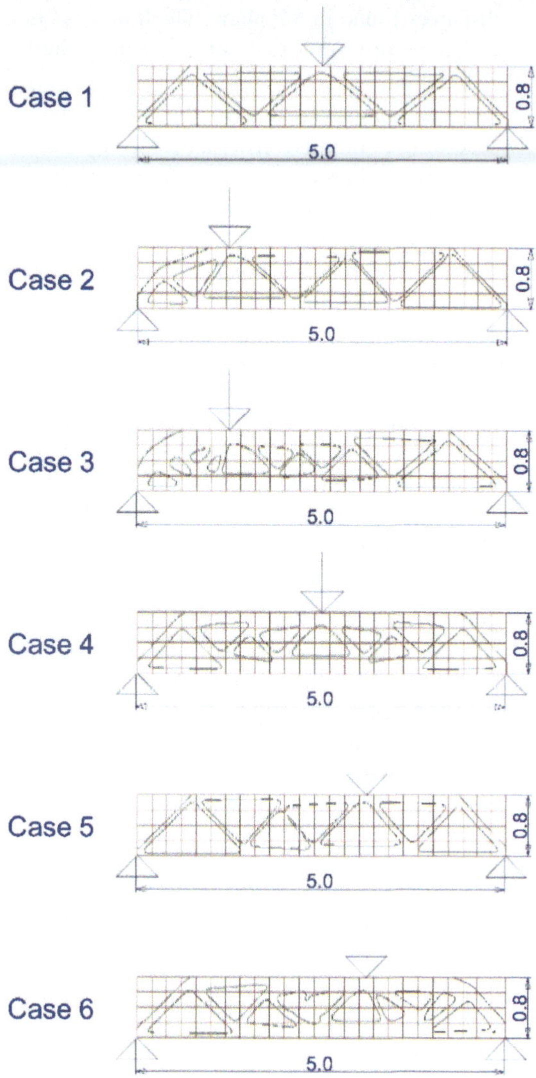

Fig. 2 The 6 case studies of Topology Optimization analysis in 2D geometrical representation. The figure illustrates overall dimensions, force, support points, cells and contours

In the case of top-down design logic, the following steps were provided:

1. The distances between cell centres and contours were calculated in *XY* plane.
2. The distances were redefined based on a simple mathematical equation in order to be reduced to values of height in *Z* direction, inversely proportional to the distances found in *XY* plane. The limit of height values was set to 0.4 m.
3. Height values were used as the starting point for developing extruded points in Z direction.
4. Extruder points were used for the 3D design of splines, polylines, or other curves according to intuitive criteria decided by each student (Fig. 3).
5. Based on different curves, free form surfaces were generated as upper surfaces using different 3D modelling approaches such as Ruled Surface and Control Point Loft (Fig. 4).
6. Additional surfaces were added to the outside boundary of the shape resulting 3D closed solid polysurfaces (Fig. 4).
7. Then, 3D negative shapes were generated by using the Difference Boolean operation.
8. Finally, both 3D solids and 3D negative shapes were mirrored in *XY* plane and the final 3D prototypes and respective 3D formworks were developed.

In bottom-up design logic, the first two steps given to top-down approach remained the same but the 3D generation of solids followed a different approach that is described as follows:

1. Height values were used as the starting point for defining 3D closed solid polysurfaces in *Z* direction that represented structural units for every cell.
2. The definition of 3D closed solids was based on intuitive criteria decided by each student. A simple example where height values were used for extruding the boundaries of individual cells in Z direction, creating 3D solid boxes with different heights is shown in Fig. 5.

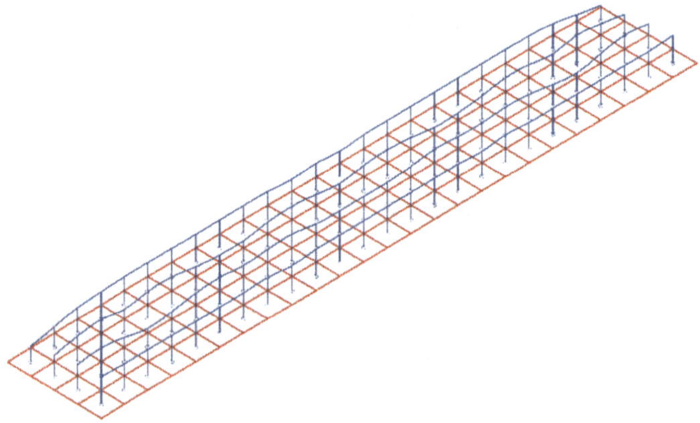

Fig. 3 An example of points extrusion and the generation of curves using input parameters of case study 1

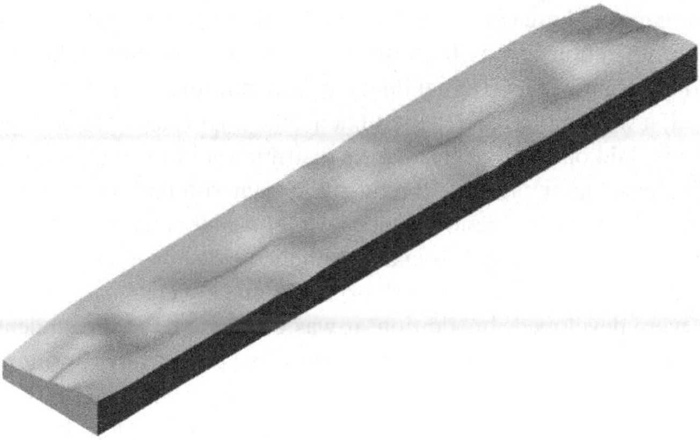

Fig. 4 An example of 3D modelling of the overall shape using input parameters of case study 1

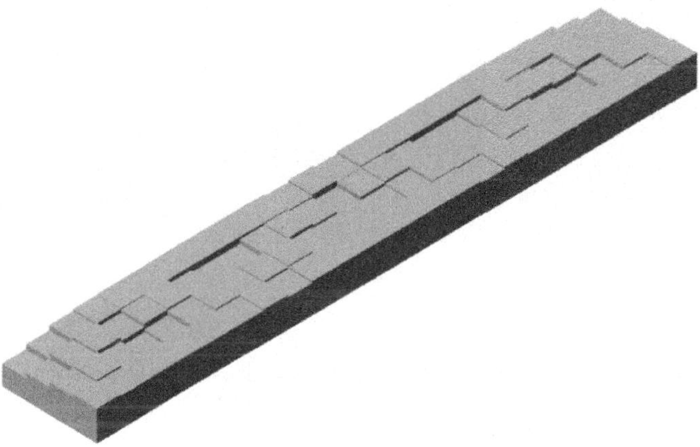

Fig. 5 An example of 3D modelling of the individual structural using input parameters of case study 1

3. The 3D closed solids were developed using different modelling approaches, both for designing and for modifying and included 3D solids and Boolean operations, respectively.
4. Following the development of 3D closed solids polysurfaces, the next step was the generation of the 3D negative shapes using Difference Boolean operation.
5. Finally, both 3D solids and 3D negative shapes were mirrored in *XY* planes and the final 3D prototypes and respective 3D formworks were developed.

Based on the abovementioned parametric design logics and after random case study distribution among the students of undergraduate course, 30 DS were developed. As it has been mentioned, the choice for each 3D shape generation was achieved based on intuitive criteria but also on constraints provided according to the case study under investigation. These constraints were related to material distribution in the 2D Topology Optimization results but also geometrical constraints related to the 2D domain and Z height of the external boundaries conditions provided.

In general, it was emphasized to the students the importance of developing design solutions that could optimally consider the distribution of the material in the generation of 3D shapes in an attempt to reduce the total volume, but at the same time, the optimum volume of 3D negative shapes was considered during the 3D modelling process. These principles were associated with the next step of the process related to digital fabrication, where 3D formwork printings and concrete casting of 1:10 scale prototypes were developed. In addition, it was emphasized to the students that the reduction of formwork material could have a positive impact on 3D printing time but also on embodied energy of the plastic material used, in this case, Polylactic Acid (PLA). Figure 6 shows design solutions generated based on top-down and Fig. 7 shows design solutions generated based on bottom-up design logic. Figure 8 shows the 3D generation of formworks.

Fig. 6 Design solutions generated based on top-down logic. The design solution (DS 2) on the left-hand side was based on input parameters of case study 4 and the design solution (DS 3) on the right-hand side was based on input parameters of case study 3

Fig. 7 Design solutions generated based on bottom-up logic. The design solution (DS 8) on the left-hand side was based on input parameters of case study 2 and the design solution (DS 28) on the right-hand side was based on input parameters of case study 5

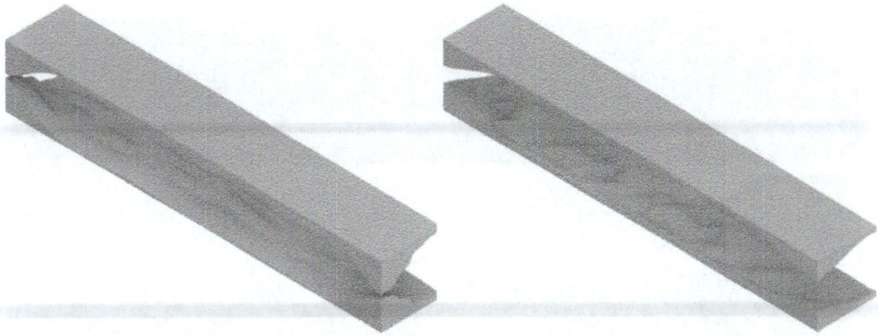

Fig. 8. 3D formwork of two design solutions. On the left-hand side the formwork for DS 2 and on the right-hand side the formwork for DS 8

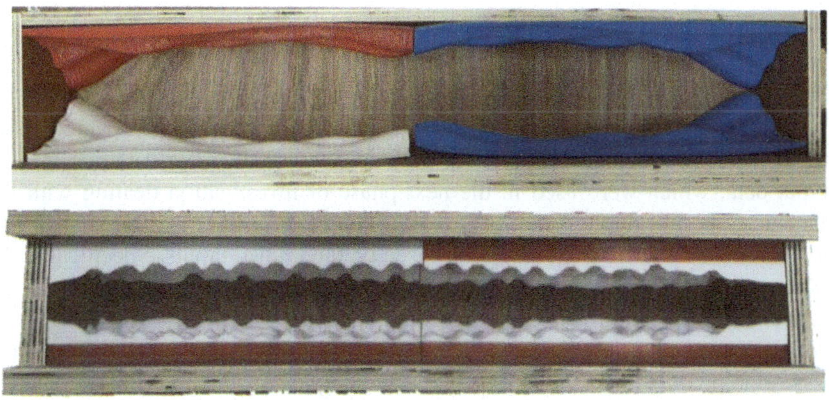

Fig. 9 Formworks structures consisting of PLA 3D printed parts. Formwork of DS 2 (top) and formwork of DS 3 (bottom)

In the next step, the fabrication process was conducted based on the derived DS, attempting to improve the sustainable aspect of the process, which included reduction of the 3D printing time of formworks but also to improve the environmental impact potential of the PLA material. The digital fabrication process involved 3D printing production of formworks in 1:10 scale based on previous design outcomes using an Ultimaker 3D printer. Each formwork was split into 4 parts due to the size limitations of the printer machine. Then, all formwork parts were assembled, creating the overall formwork structure that was ready for concrete casting. Figure 9 shows representative formwork structures and Fig. 10 shows physical prototypes of DS 3 and DS 8.

Fig. 10 Physical prototypes. Physical prototype of DS 3 (top) and physical prototype of DS 8 (bottom)

3.3 Post-processing analysis of DS

The DS obtained during the digital design to fabrication development were further processed using Grasshopper (plug-in for Rhino software) in order to extract specific output data, which were used in the next phase of investigation dealing with the Machine Learning implementation for the classification of existing DS, but also the prediction of new design solutions based on the trained inputs. Specifically, the DS were distinguished according to the input parameters involved using the 6 case studies. Then, the training inputs included Force point position (%), Volume fraction (VolFrac), and Filter size (Rmin). Based on the results derived from digital design to fabrication process of the 30 DS, the expected outputs included Formwork volume (m^3), Formwork printing time (min), Filament length (mm), Material cost (Euro), and Embodied Carbon ($KgCO_2e/Kg$). The cost for the PLA material was calculated as 40 Euro/Kg. The Embodied Carbon was calculated based on Cardinal LCA plug-in [23] in Grasshopper (plug-in for Rhino). Specifically, using cradle-to-gate procedure a custom material as PLA was developed by specifying the Global Warming Potential ($KgCO_2e/Kg$) for stages A1–A3 at 0.501 $KgCO_2e/Kg$ and the Density (Kg/m^3) at 1210 kg/m^3 based on existing works in LCA analysis of PLA material [24, 25]. The Embodied carbon results for each DS are derived by multiplying the formwork volume in 1:10 scale with the Global Warming Potential and the Density of material. The input and output data are summarized in Table 2.

3.4 Machine Learning (ML) implementation

The results derived from post-processing investigation were used as training input and expected output data in a supervised neural network algorithm, which classify and predict the DS of complex beam structures with specific loading conditions and

Table 2 Training inputs and expected outputs of the 30 DS

Design solutions (DS)	Cases	Training inputs			Expected outputs					
		Force point position (0–1)	Volume fraction (VolFrac)	Filter size (Rmin)	Formwork volume (m^3)	Formwork printing time (min)	Filament length (mm)	Plastic weight (g)	Material cost (Euro)	Embodied carbon ($kgCO_2$ e/kg)
DS_01	1	0.50	0.40	2.00	1.00	1268	160,565.8	482.76	19.31	0.00060347
DS_02	4	0.50	0.60	1.50	0.70	1211	133,475.1	401.31	16.06	0.000420037
DS_03	3	0.25	0.60	1.50	1.81	2448	268,452.7	807.13	32.28	0.001099241
DS_04	5	0.62	0.40	2.00	0.76	1463	147,106.1	442.29	17.69	0.000410113
DS_05	2	0.25	0.40	2.00	1.04	2785	226,305	680.4	27.22	0.00063252
DS_06	3	0.25	0.60	1.50	1.77	3333	308,842.6	928.57	37.14	0.001073174
DS_07	1	0.50	0.40	2.00	2.07	2697	297,430.9	894.26	35.77	0.001253703
DS_08	2	0.25	0.40	2.00	0.49	1419	128,526	386.43	15.46	0.00029691
DS_09	6	0.62	0.60	1.50	1.49	2529	256,999.5	772.69	30.91	0.000863607
DS_10	5	0.62	0.40	2.00	0.68	1307	135,125.9	406.27	16.25	0.00041472
DS_11	1	0.50	0.40	2.00	0.67	1265	133,126.4	400.26	16.01	0.000311349
DS_12	1	0.50	0.40	2.00	1.70	2850	275,453.2	828.18	33.13	0.001031042
DS_13	6	0.62	0.60	1.50	1.65	3186	297,420.3	894.23	35.77	0.000998246
DS_14	2	0.25	0.40	2.00	1.27	2335	215,899.7	649.13	25.96	0.000768674
DS_15	2	0.25	0.40	2.00	0.85	1696	161,666.4	486.07	19.45	0.000517467
DS_16	1	0.50	0.40	2.00	1.47	2822	276,474.5	831.25	33.25	0.000890098
DS_17	3	0.25	0.60	1.50	1.40	2237	234,955.8	706.42	28.25	0.001262735
D5_18	4	0.50	0.60	1.50	1.79	2708	276,136.8	830.23	33.21	0.001082145
DS_19	5	0.62	0.40	2.00	2.35	4021	382,971.8	1151.44	46.06	0.001426048

(continued)

Table 2 (continued)

Design solutions (DS)	Cases	Training inputs			Expected outputs					
		Force point position (0–1)	Volume fraction (VolFrac)	Filter size (Rmin)	Formwork volume (m^3)	Formwork printing time (min)	Filament length (mm)	Plastic weight (g)	Material cost (Euro)	Embodied carbon ($kgCO_2$ e/kg)
DS_20	6	0.62	0.60	1.50	1.12	2764	279,954	841.71	33.67	0,000,676,348
DS_21	3	0.25	0.60	1.50	1.88	4081	331,352.3	996.24	39.85	0.001139917
DS_22	2	0.25	0.40	2.00	2.83	4333	426,518.9	1282.37	51.3	0.001715635
DS_23	4	0.50	0.60	1.50	1.00	2019	186,320.2	560.19	22.41	0.000606137
DS_24	4	0.50	0.60	1.50	0.77	1478	158,965.2	477.94	19.12	0.000466479
DS_25	1	0.50	0.40	2.00	1.56	2586	252,256.5	758.44	30.34	0.000932775
DS_26	1	0.50	0.40	2.00	1.04	1689	183,140	550.63	22.02	0.000632944
DS_27	5	0.62	0.40	2.00	1.58	2586	257,815.9	775.15	31.01	0.00095375
DS_28	5	0.62	0.40	2.00	1.09	1894	196,946.4	592.14	23.69	0.000661618
DS_29	4	0.50	0.60	1.50	1.02	2596	240,537.5	723.2	28.92	0.000616879
DS_30	6	0.62	0.60	1.50	1.15	2031	195,997.7	589.29	23.58	0.000698051

geometrical inputs. Specifically, the 3 training inputs and the 6 expected outputs from the 30 DS along with their corresponding case studies (Table 2) were used during the training process. The Artificial Neural Network (ANN) model is developed using LunchBox plug-in [26] which is an ML tool for Grasshopper (plug-in for Rhino software). Initially, the data derived from the design solutions was classified in order to separate inputs from outputs. Therefore, this data was remapped to numbers from 0 to 1 and fed into a backpropagation algorithm with a sigmoid activation function. Further to this, the train model includes two hidden layers with 4 neurons on each layer. The number of hidden neurons has been chosen through a trial and error process, in order to find the most precise prediction possible.

Figure 11 describes in detail the basic steps of the process with the integration of the ANN algorithm. The diagram indicates the testing process, where according to the input from each case, it predicts 6 outputs. In this process, 100,000 iterations were performed with error values from 0.000409 to 0.00005 (Fig. 12). The ANN model allows a comparison of the predicted outputs with the expected outputs from the design solutions of each case. For example, a case input can indicate the predicted embodied carbon which is compared with the expected embodied carbon from the design solutions of that certain case. The overall outcome of this comparison is analytically presented in the next section.

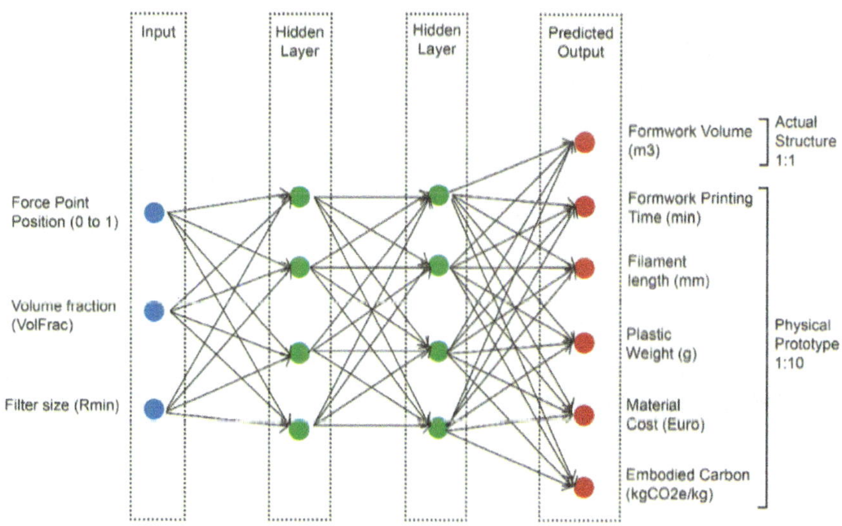

Fig. 11 ANN diagram describing inputs and predicted outputs

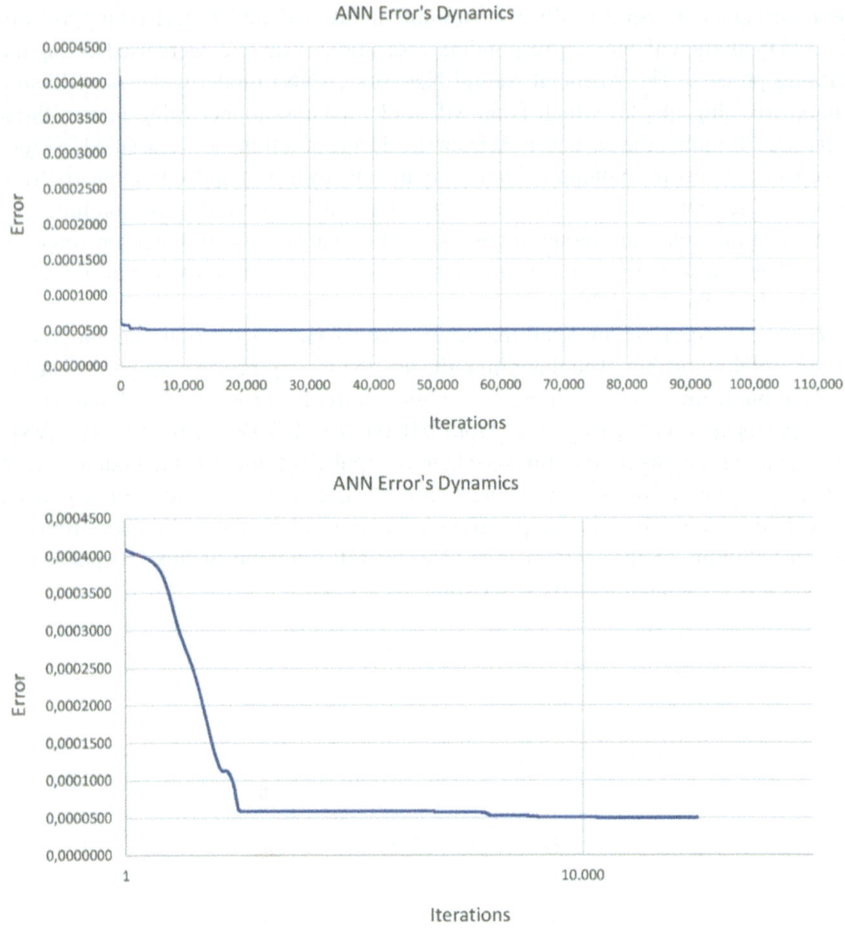

Fig. 12 ANN diagram describing Error's Dynamics

4 Results and discussion

This section provides detailed results regarding classification and prediction of existing DS, which were produced during the previous phase of experimentation but also results that were derived by testing New Design Solutions (NDS) based on the same ANN. Specifically, in the first part, the results are classified based on predicted outcomes derived from the 30 DS and are distinguished according to the 6 case studies, considering the expected performance outputs. In the second part, the performance of 6 NDS is examined, using new input data based on a combination of existing and new input parameters.

4.1 *Classification and prediction of Design Solutions (DS)*

Although the ML algorithms implementation requires a large number of design solutions during the training phase so that the predicted results are as accurate as possible and respond to the expected outputs without major deviations, by exploring the first 30 DS, a first insight into the capabilities of the ANN can be provided. Within this framework, the classification of the DS based on the input parameters of case studies but also the expected outcomes is presented in the following figures. This achieves a comparison between expected and predicted outputs in each case study.

Figure 13 demonstrates predicted and expected results for DS, where input parameters of case study 1 were applied. The DS examined were DS 01, DS 07, DS 11, DS 12, DS 16, DS 25, and DS 26. In this figure, 6 different graphs represent each of the 6 different output values derived from the experimentation; Formwork volume (m^3), Formwork printing time (min), Filament length (mm), Plastic weight (g), Material cost (Euro) and Embodied carbon ($KgCO_2e/Kg$). As it can be observed, the average expected value for each output is close to the predicted result derived through the ANN. Also, small deviations between ANN prediction and average output results of

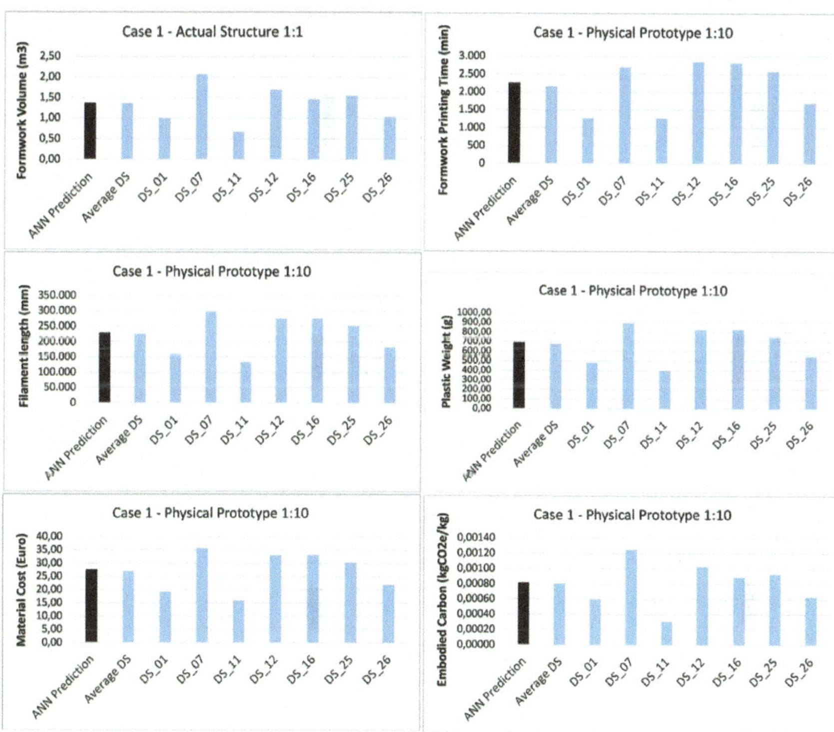

Fig. 13 Graphs that represent predicted and expected results of the 6 outputs using input parameters of case study 1. Comparisons between ANN prediction and average outputs are provided

DS using input parameters of case study 1 were observed. Specifically, the comparison between the ANN prediction value and the average expected value in the case of Formwork volume (m^3) shows deviation at a percentage rate of 0.70%. The comparison between the predicted and the average expected values in case of Filament length (mm), Plastic weight (g), Material cost (Euro) and Embodied carbon ($KgCO_2e/Kg$) indicates deviation at a percentage rate that range between 1.73 and 2.29%.

Similarly, Fig. 14 demonstrates predicted and expected output results for DS, where input parameters of case study 2 were applied. The DS examined were DS 05, DS 08, DS 14, DS 15, and DS 22. In this case, the comparison between the ANN prediction value and the average expected value in the case of Formwork volume (m^3) shows deviation at a percentage rate of 4.31%. The comparison between the predicted and the average expected values in the case of Filament length (mm), Plastic weight (g), Material cost (Euro), and Embodied carbon ($KgCO_2e/Kg$) indicates deviation at a percentage rate that ranges between 4.32 and 5.22%.

The DS examined in case study 3 were DS 03, DS 06, DS 17, and DS 21. In case study 4 the DS 02, DS 18, DS 23, DS 24, and DS 29. In case study 5 the DS 04, DS 10, DS 19, DS 27, and DS 28. Finally, in case study 6 the DS 09, DS 13, DS 20,

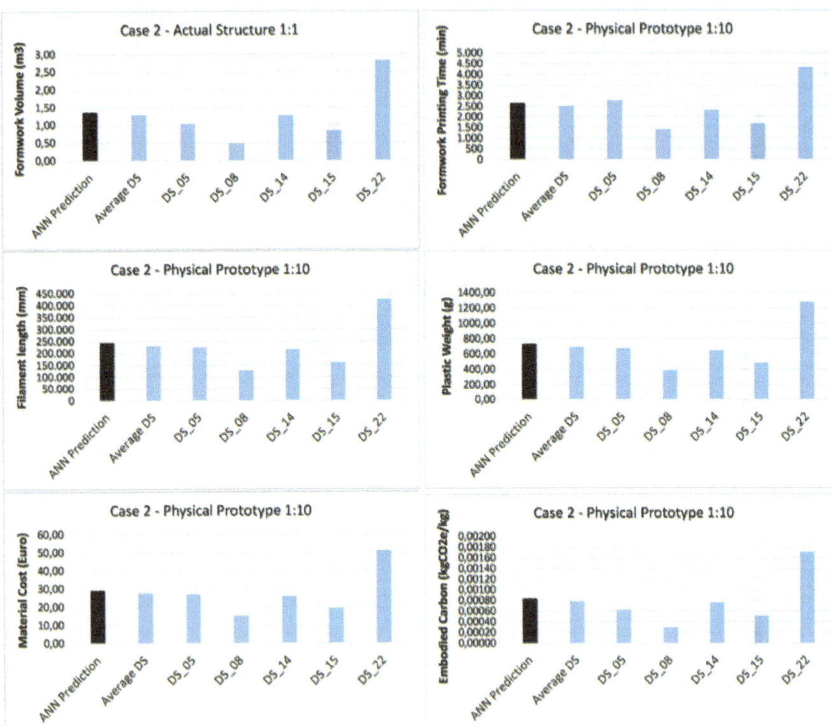

Fig. 14 Graphs that represent predicted and expected results of the 6 outputs using input parameters of case study 2. Comparisons between ANN prediction and average outputs are provided

and DS 30 were investigated. In all case studies, the ANN prediction values and the average expected values for all 6 performance outputs were compared. In case study 3 the comparison of the 6 outputs indicates deviation at a percentage rate that range between 3.11 and 3.87%. In case study 4 deviations at a percentage rate that range between 0.16 and 1.50% are observed. In case study 5 the percentage rate ranges between 1.29 and 6.34%, while in case study 6 deviations at a percentage ranging between 1.81 and 2.53% are observed.

The detailed investigation of the output results obtained as regards the sustainable potential of the DS show that the ANN prediction value of Material cost (Euro) in case study 4 is the minimum whereas in the case study 3 is the maximum (Fig. 15). Also, the ANN prediction value of Embodied carbon ($KgCO_2e/Kg$) in case study 4 is the minimum, whereas in the case study 3 is the maximum (Fig. 16).

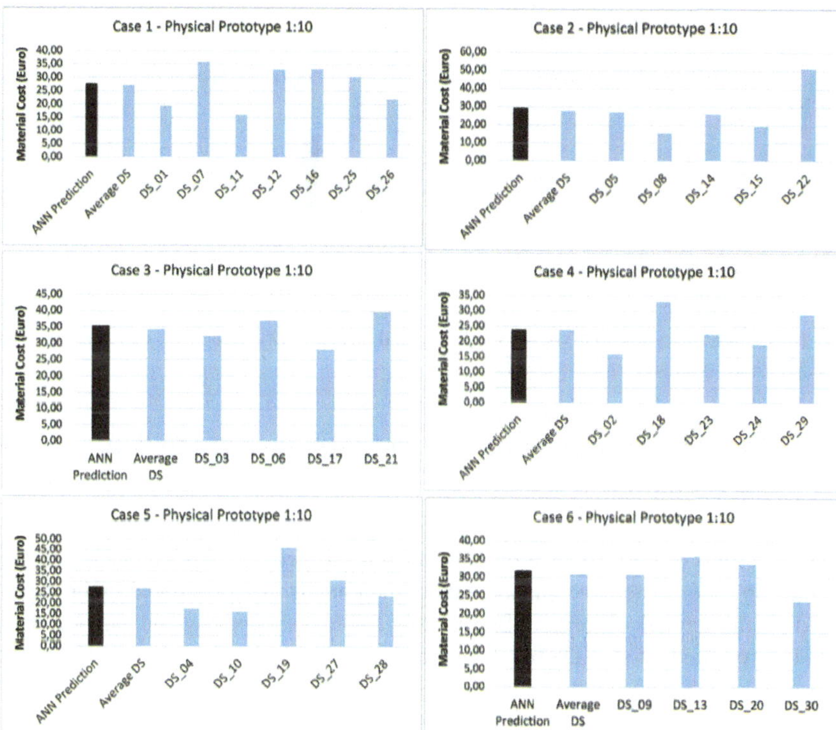

Fig. 15 Graphs that represent ANN predicted and average DS values of material cost (Euro) for all case studies (1–6)

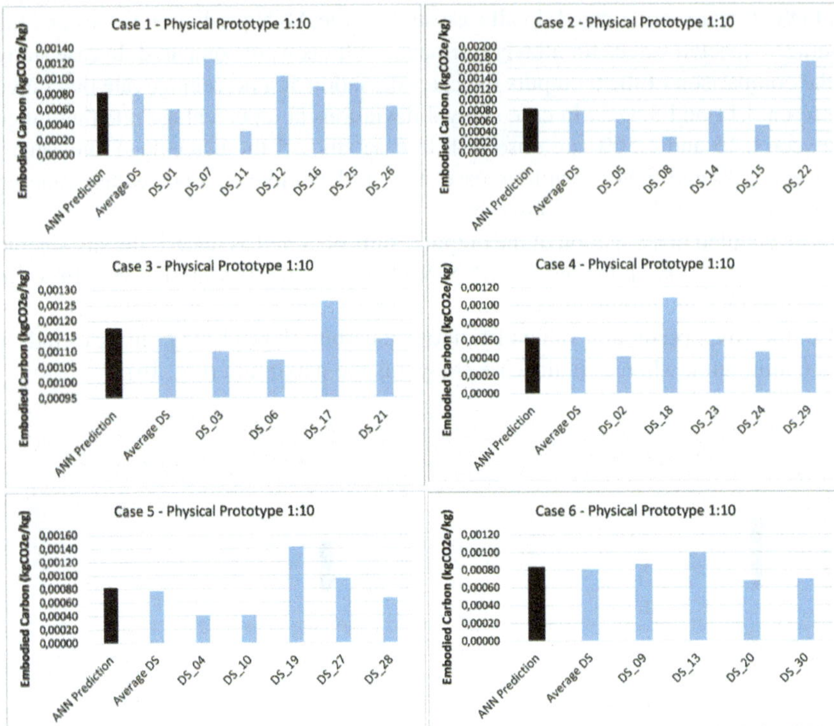

Fig. 16 Graphs that represent ANN predicted and average DS values of embodied carbon ($KgCO_2e/Kg$) for all case studies (1–6)

4.2 New Design Solutions (NDS) performance prediction and validation

The prediction of performance outputs derived through ANN is conducted in this section. In order to achieve this, new input parameters are formulated in the supervised neural network algorithm. Specifically, Table 3 shows new and existing input data incorporated into 6 NDS based on the 6 case studies developed for TO analysis. The first 3 NDS incorporate 1 new and 2 existing input parameters while the last 3 NDS incorporate 2 new and 1 existing parameters.

Predicted results of the 6 NDS solutions for 6 performance outputs; Formwork volume (m^3), Formwork printing time (min), Filament length (mm), Plastic weight (g), Material cost (Euro) and Embodied carbon ($KgCO_2e/Kg$) are extracted from ANN. Analytically, Fig. 17 demonstrates the predicted performance output results for all NDS. As it can be generally observed, the values of NDS 01, NDS 03, NDS 05, and NDS 06 in all predicted performance outputs appear to be reduced relative to the values of NDS 02 and NDS 04.

Table 3 Input data consisting of 6 NDS used in ANN for output results prediction

New design solutions (NDS)	Force point position (0–1)	Volume fraction (VolFrac)	Filter size (Rmin)
1	0.75 (new)	0.4 (existing)	2 (existing)
2	0.25 (existing)	0.75 (new)	1.5 (existing)
3	0.62 (existing)	0.6 (existing)	1.75 (new)
4	0.12 (new)	0.6 (existing)	1.6 (new)
5	0.5 (existing)	0.45 (new)	1.8 (new)
6	0.86 (new)	0.55 (new)	2 (existing)

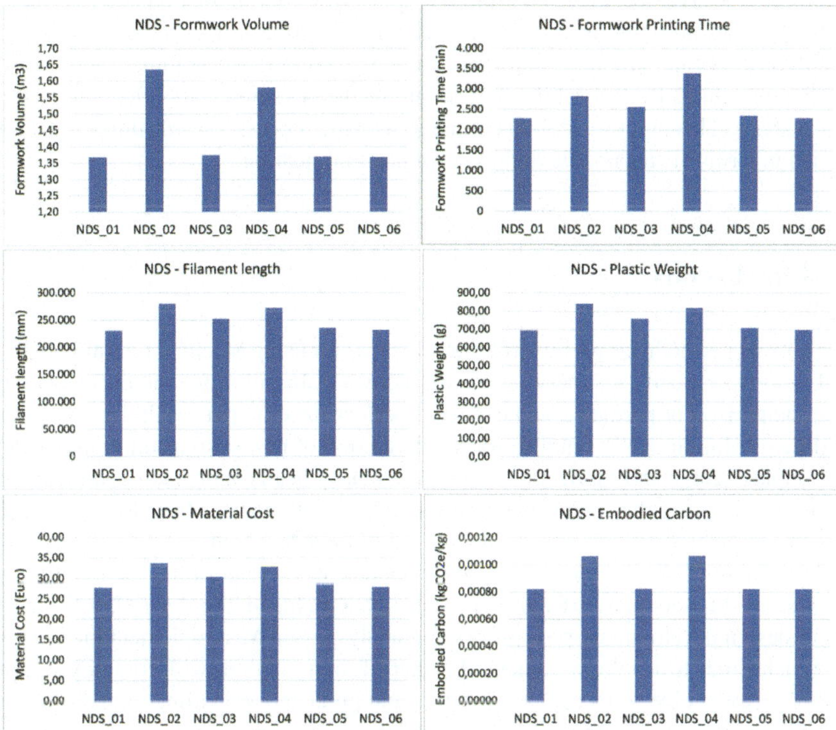

Fig. 17 Graphs that represent the results of the 6 predicted performance outputs in the case of 6 NDS

The comparison between the NDS where reduced values are observed shows that in the case of Formwork volume (m^3), the best minimum result is the same for the 4 NDS out of 6 (NDS 01, NDS 03, NDS 05, and NDS 06). In the case of Formwork printing time (min), the best minimum result is observed in NDS 01. The same applies in the cases of Filament length (mm), Plastic weight (g), and Material cost (Euro), where the best minimum values are observed in NDS 01. In the case of Embodied carbon (KgCO2e/Kg), where the calculation is related to the Formwork volume (m^3), 4 NDS have almost the same minimum value.

In general, the comparison between the performance output results of all NDS show that NDS 01 achieves the best fitting results in all categories with the cases NDS 03, NDS 05, and NDS 06 to follow with slight differences in each output case. If the output performance is examined in relation to the input values, the input data of NDS 01 refer to 0.75 (0–1) for Force point position, 0.4 (VolFrac) for Volume fraction, and 2 (Rmin) for Filter size. A comparison between the input values of 4 NDS best optimum solutions shows that in the case of Force point position best results are observed between 0.5 and 0.88 (0–1). In the case of Volume fraction best results are observed between 0.4 and 0.6 (VolFrac) and in the case of Filter size best results are observed between 1.75 and 2 (Rmin). Considering that NDS 01 is found to be the optimum one, input selection of force position in the 3/4 of the overall length of the beam structure, minimum value 0.4 VolFrac for Volume fraction and maximum value 2 Rmin for Filter size might be the best input data selection during TO analysis. This input selection is found to lead the process towards the best fitting results in terms of sustainable digital design and fabrication.

5 Conclusions

The current research has presented a methodology for using Machine Learning (ML) and more specifically a supervised neural network algorithm for the classification and comparison of predicted and expected performance output results in a number of design solutions (DS), but also for the prediction of new design solutions (NDS) in complex design and fabrication tasks, which have been conducted by students of architecture within the framework of an undergraduate course. In order to achieve this, a series of data has been extracted and applied, which has formulated the training inputs and the expected outputs applied in the algorithm. This data has been derived based on a series of approaches that included Topology Optimization (TO), design development, post-processing analysis, and ML implementation. This approach aimed to develop a framework, where design solutions with complex design and fabrication objectives can be classified and predicted, providing the ground for specifying inputs and outputs that are best fitting results and can lead to sustainable outcomes in terms of material usage and environmental impact.

The two categories of results obtained through the classification and prediction of Design Solutions (DS) and the New Design Solutions (NDS) performance prediction and validation show promising directions for implementing ML in the digital design and fabrication process. The results obtained showed that, although design and fabrication process is a complex task that involves a large number of intuitive decisions taking during the development of structures and products, the classification and prediction of best fitting solutions can be achieved. More specifically, in the first part of the investigation the 30 DS have been classified and ANN predicted results have been derived and compared with the average expected results in each case study examined. The results show small deviations, if the comparison between predicted and expected results is examined, showing the algorithm might be able to predict future design and fabrication outcomes. In the second part of the investigation, the application of 6 NDS and the results obtained using the same supervised neural network algorithm show that best fitting design solutions can be identified based on input data in an attempt to predict the design and fabrication of physical prototypes before actual work is conducted. Also, the selection of input data that can respond to best fitting solutions is feasible and can provide specific input information that can be used during Topology Optimization analysis.

This research has also demonstrated possible directions for further improvements. The use of 30 DS is considered to be a small sample in ML investigation process. Future work towards the increase of design solutions is considered an important aspect of ML investigation. Also, the incorporation of 3 training inputs and 6 predicted outputs should be extended by indicating new training inputs and predicted outputs in the algorithm in an attempt to include qualitative criteria involved during design and fabrication stages. The ultimate aim is to classify architectural solutions based on their sustainable performance and in parallel to raise awareness among the students of architecture about digital design and fabrication workflows that can have a positive impact on the built environment.

Acknowledgements The design solutions presented and analysed in this paper were developed within the framework of the undergraduate course ARH-220 Digital Architectural Communication Media taught by Odysseas Kontovourkis, Ph.D., in the Department of Architecture at the University of Cyprus during the Fall semester of 2019. We would like to sincerely thank all the students who attended the course and their works are presented in this paper.

References

1. De Schutter G, Lesage K, Mechtcherine V et al (2018) Vision of 3D printing with concrete—technical, economic and environmental potentials. Cem Concr Res 112:25–36
2. Buswell RA, Soar R, Pendlebury M, et al (2005) Investigation of the potential for applying freeform processes to construction. In: Proceedings of the 3rd international conference on innovation in architecture, engineering and construction (AEC) 2005. Rotterdam, The Netherlands, pp 141–150
3. Sivanandam S, Deepa S (2007) Introduction to genetic algorithms, 1st edn. Springer, Berlin, Heidelberg
4. Saldana Ochoa K, Ohlbrock P, D'Acunto P, Moosavi V (2020) Beyond typologies, beyond optimization: exploring novel structural forms at the interface of human and machine intelligence. Int J Archit Comput. https://doi.org/10.1177/1478077120943062
5. Tamke M, Nicholas P, Zwierzycki M (2018) Machine learning for architectural design: practices and infrastructure. Int J Archit Comput 16:123–143. https://doi.org/10.1177/147807711877 8580
6. Thomsen MR, Nicholas P, Tamke M, et al (2019) Predicting and steering performance in architectural materials. In: Architecture in the age of the 4th industrial revolution: proceedings of the 37th eCAADe and 23rd SIGraDi conference. pp 485–494

7. Thomsen MR, Nicholas P, Tamke M et al (2020) Towards machine learning for architectural fabrication in the age of industry 4.0. Int J Archit Comput 18:335–352. https://doi.org/10.1177/1478077120948000
8. Rebala G, Ravi A, Churiwala S (2019) An introduction to machine learning. Springer Nature Switzerland AG, Cham
9. Oppermann A (2019) What is deep learning and how does it work? https://towardsdatascience.com/what-is-deep-learning-and-how-does-it-work-2ce44bb692ac. Accessed 15 May 2021
10. Peng W, Zhang F, Nagakura T (2017) Machines' perception of space: employing 3D Isovist methods and a convolutional neural network in architectural space classification. In: Disciplines and disruption—Proceedings Catalog of the 37th Annual Conference of the Association for Computer Aided Design in Architecture, ACADIA 2017. pp 474–481
11. Nicholas P, Rossi G, Williams E et al (2020) Integrating real-time multi-resolution scanning and machine learning for Conformal Robotic 3D Printing in Architecture. Int J Archit Comput 18:371–384. https://doi.org/10.1177/1478077120948203
12. Nicholas P, Zwierzycki M, Nørgaard EC, et al (2017) Adaptive robotic fabrication for conditions of material inconsistency increasing the geometric accuracy of incrementally formed metal panels. In: Fabricate 2017. UCL Press, London, pp 114–121
13. Rossi G, Nicholas P (2018) Modelling a complex fabrication system: New design tools for doubly curved metal surfaces fabricated using the English Wheel. In: 36th eCAADe 2018: Computing for a better tomorrow. Lodz, Poland, pp 811–820
14. Sanatani RP (2020) A machine-learning driven design assistance framework for the affective analysis of spatial enclosures. RE Anthr Des Age Humans—Proc 25th Int Conf Comput Archit Des Res Asia, CAADRIA 2020 1:741–750
15. Rossi G, Nicholas P (2018) Re/learning the wheel: methods to utilize neural networks as design tools for doubly curved metal surfaces. In: Proceedings of the ACADIA 2018: Re/calibration: on imprecision and infidelity. Mexico City, Mexico, pp 146–155
16. TopOpt. https://www.food4rhino.com/en/app/topopt. Accessed 22 Dec 2021
17. Grasshopper. https://www.grasshopper3d.com/. Accessed 22 Dec 2021
18. Rhinoceros. https://www.rhino3d.com/. Accessed 22 Dec 2021
19. Sigmund O (2001) A 99 line topology optimization code written in Matlab. Struct Multidiscip Optim 21:120–127
20. Li D, Liao W, Dai N et al (2018) Optimal design and modeling of gyroid-based functionally graded cellular structures for additive manufacturing. Comput Des 104:87–99
21. Cheng L, Bai J, To AC (2019) Functionally graded lattice structure topology optimization for the design of additive manufactured components with stress constraints. Comput Methods AppliedMechanics Eng 344:334–359
22. Tang Y, Kurtz A, Zhao YF (2015) Bidirectional evolutionary structural optimization (BESO) based design method for lattice structure to be fabricated by additive manufacturing. Comput Des 69:91–101
23. Cardinal LCA. https://www.food4rhino.com/en/app/cardinal-lca. Accessed 22 Dec 2021
24. Morão A, de Bie F (2019) Life cycle impact assessment of polylactic acid (PLA) produced from sugarcane in Thailand. J Polym Environ 27:2523–2539
25. Rezvani Ghomi E, Khosravi F, Saedi Ardahaei A et al (1854) (2021) The life cycle assessment for polylactic acid (PLA) to make it a low-carbon Material. Polym 2021:13
26. LunchBox. https://www.food4rhino.com/en/app/lunchbox. Accessed 22 Dec 2021

Evaluation of the Station Uniform for Firefighters by Anthropometric Method on the Basis of the Physico-mechanical Properties of the Material

Ada Traumann, Teele Peets, Merje Beilmann, and Gertu Vilba

The aim of the research was to test firefighters' clothing in different body positions during their work using the anthropometric method developed at TTK University of Applied Sciences, according to which the suitability of the size number and height of a firefighter's station uniform can be assessed. Research conducted with the Rescue Board revealed that more complex body positions test the properties of special clothing materials and cause discomfort to the wearer. To compensate for the little elongation of the garment material, firefighters in many cases chose clothes one size larger to add some extra space. The anthropometric method includes 12 different movements, carefully selected from the body positions encountered in firefighters' tasks. 61 firefighters who evaluated the special clothing took part in the testing. The evaluation was performed by four different teams across Estonia. However, new problems were added in the evaluation of garments with this method, the firefighters who participated in the experiment complained about the inappropriate composition of the special garment materials, their elongation and permeability of fabrics. The station uniform of firefighters consists of a T-shirt and uniform trousers as the garment assembly, which are worn daily 24/7 depending on the work tasks, also during rescue work with protective clothing. In the case of special clothing, it is important to find a suitable fabric that, with its fibrous composition and structure, helps to reduce body heat without creating a potential risk of heat stress and reducing the firefighter's performance in work operations, which can be life-threatening in certain situations.

A. Traumann (✉) · T. Peets · M. Beilmann · G. Vilba
Institute of Engineering and Circular Economy, TTK University of Applied Sciences, Parnu Road 62, 10135 Tallinn, Estonia
e-mail: ada.traumann@tktk.ee

T. Peets
e-mail: teele.peets@tktk.ee

M. Beilmann
e-mail: merje.beilmann@tktk.ee

P. Kyratsis et al. (eds.), *Advances in Product Design Engineering*, Management and Industrial Engineering,
https://doi.org/10.1007/978-3-030-98124-2_6

This research revealed that the biggest problem for firefighters was the material of the uniforms, which has low air permeability, makes the wearer sweat and sometimes obstructs movement. A wet T-shirt is uncomfortable for the wearer and increases the risk of cold.

Based on the information obtained by testing the station uniform of firefighters by the anthropometric method, air permeability and elongation tests have been performed on the fabric of the firefighters' T-shirt with a polyester mixture. The obtained results have been compared with different T-shirt fabrics and a test application has been performed on several fibrous fabrics.

1 Introduction

1.1 Static Anthropometry

The term "anthropometry" is derived from a combination of the Greek words *anthrop* (meaning man) and *metricos* (meaning measurement). It refers to the systematic measurement and collection of data on human physical properties such as body dimensions, body shape, masses of body segments, centre of gravity and inertial properties [1].

Nowadays, anthropometry plays an important role in engineering, as part of industrial design, clothing design and architecture, where these data are used in the design and evaluation of user-centred systems, equipment, industrially manufactured products, workspaces and other facilities [1].

Static or structural anthropometry is the traditional and simplest method for collecting anthropometric data. Circumferential dimensions, lengths and widths of body parts are recorded by holding the body in fixed standard positions. [1] Methods for collecting such measurements are outlined by Bye et al. 2006, who classified them as follows [2]:

- Linear measurements, which provide data from the distance between two points.
- A multi-probe method using a combination of linear methods with other tools to map body contours.
- The body shape method, which provides information about the surface and shape of the body.

In order to determine the size number of the Estonian Rescue Board's clothing, the wearers need to know or re-measure at least 11 different body measurements. Figures 1 and 2 show the most important body measurements for both women and men, which are height (A), head circumference (B), neck at base girth (C), bust/chest girth horizontal (D), arm length (E), waist girth (F), buttock girth (G), inseam (H), foot width and foot length (I), palm length (J), palm girth (K).

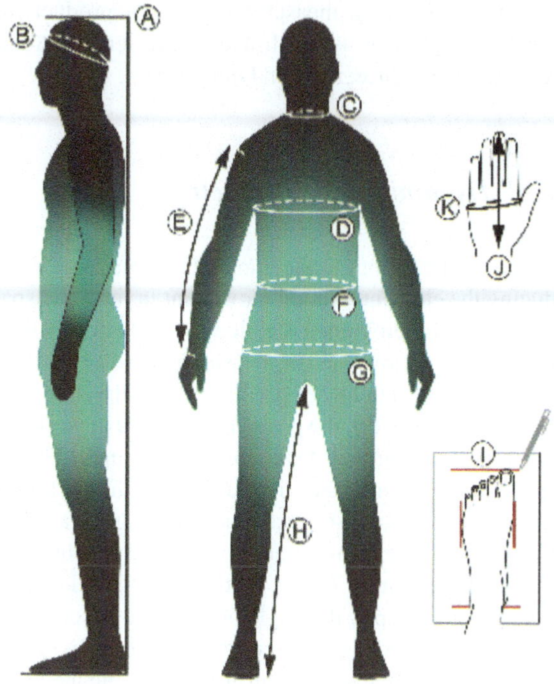

Fig. 1 Men's body measurements

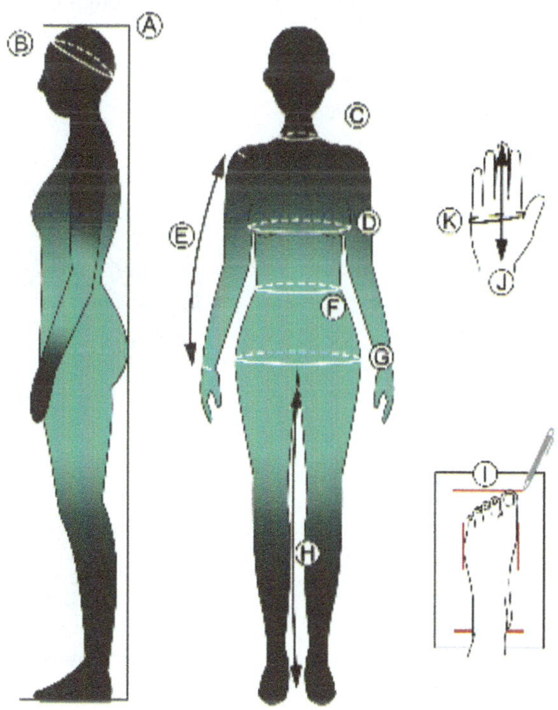

Fig. 2 Women's body measurements

The main sizing measurements for a product that rests on the shoulder are height, bust and waist girth and sleeve length. Additional measurements include head circumference, neck and hip girth.

1.2 Dynamic Anthropometry

When wearing clothes, the human body and the clothes influence each other. Introducing the element of body movement in time and space adds a fourth dimension to the field of anthropometry, called dynamic or functional anthropometry. Functional anthropometry involves the measurement of a body in motion, or at a moment in time when various actions are performed. Clothing design is often task-specific. When designing workwear, the general movement of workers and the positions they assume when performing their tasks can be examined. The process involves observing the movement of joints and muscles during the performance of a task, evaluating the measurements against those that will be generated in actual use. If special equipment is used, the measurements must take account of additional equipment (helmet, mask, oxygen tank). This is followed by the selection of different ranges of motion, where, for example, the angle of the knee is set at 60° or 90° when monitoring the joints. The test subjects try several variations with the product being tested. When analysing the photos taken during the test, the method can be used in which notional intersecting surfaces (Fig. 3) are drawn through fixed points on the body [1].

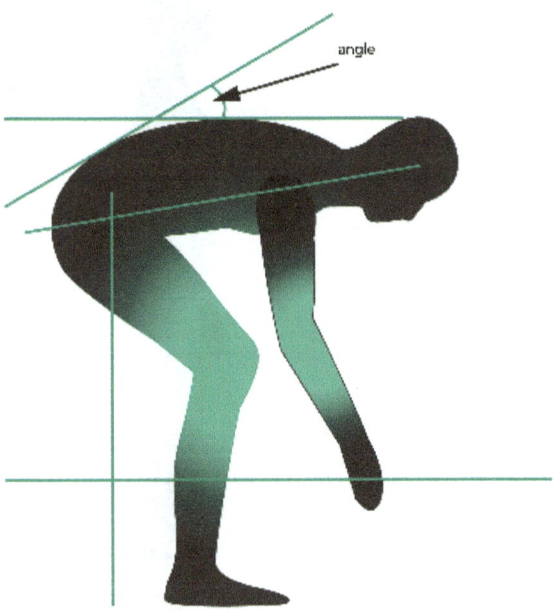

Fig. 3 Determining dimensions using dynamic anthropometry

Dynamic anthropometry is difficult and cumbersome to perform, but can be applied directly in the design process. The most successful application of dynamic methods so far has been in the design of sportswear for top athletes. The use of such data in the design of swimwear and other sportswear has been shown to reduce muscle fatigue, increase comfort, reduce resistance and in turn lead to improved performance. Of course, it is argued that in general it is not possible to measure body movement with the same accuracy as in the static position, as the course of action varies when the task is repeated, even for the same individual. It is also difficult to track the markings on the body while the body is moving. Although dynamic anthropometry adds much accuracy and versatility to the design process, today it is still a rather time-consuming and costly process [1].

1.3 Methodology for Anthropometric Measurements

Clothes form a layer close to the human body, creating an intimate environment around it. People routinely move around and perform a variety of tasks in their personal and professional lives. In the course of these activities, clothing moves and interacts with the body. A number of factors determine a pleasant or unpleasant user experience. Among these, how the garment fits or how it matches the wearer's size and body shape is an important factor. The fit of the garment affects both the appearance and performance of the wearer [1].

The basic anthropometric measurements in the technological design of garments are presented in the first part of the international standard ISO 7250-1:2017, which focuses on the definitions of body measurement and its reference points. The standard document includes a description of anthropometric measurements that can be used as a basis for comparisons between population groups and for the creation of anthropometric databases. A basic list of anthropometric measurements is defined in this standard document to assist ergonomists who need to define population groups and apply their knowledge to the geometrical design of people's living and working environments. In addition, the list is the basis for extracting one- and two-dimensional measurements from three-dimensional scans (specified in ISO 20685). It is a guide to anthropometric measurements, but in turn provides information for the ergonomist and designer on the anatomical and anthropometric bases and principles of measurement used in the design tasks [3].

The ISO 7250-1:2017 standard sets out four conditions under which measurements should be taken [3]:

1. Clothing of the subject—at the time of measurement, the subject should wear minimal clothing, be bareheaded and have no footwear.
2. Support surface—floors, support surfaces and seating areas should be flat, horizontal and non-deforming.

3. Body symmetry—both sides of the body should be equally measured. If it is not possible to measure both sides of the person, it should be noted which side is measured.
4. Body position—when standing, the person should look straight ahead, stand with heels side by side, maintain a normal posture and breathe at a normal rhythm; when seated, the torso should be upright, the shoulders relaxed, the person looks straight ahead (Frankfurt plane) and their legs are supported so that the femurs are horizontal and parallel to each other. An adjustable footrest may also be used to fix the position of the legs or a combination of different platforms may be used to achieve the desired position.

2 Ergonomics of the Work Environment

2.1 Occupational Health and Safety in Specialised Services

The firefighter profession is a dangerous one, involving intense physical activity in hazardous environments [4]. Over the years, the nature of firefighting and the role of firefighters have changed and, as a result, their equipment has changed. Today, in addition to extinguishing fires, the work of firefighters also involves rescuing people; responding to emergency incidents and car accidents, and coming into contact with hazardous substances. In addition, fires are hotter and more dangerous due to the new building materials used in today's building structures [5].

In the case of firefighters, it is important to note that, in addition to the correct working practices, the clothing worn for specific tasks also has an impact on the wearer's state of health. The wearer must be able to sense their body well and react to situations where they feel that performance is declining. If there is a noticeable decrease in physical performance, this is already a sign of a reduced state of health. The international standard ISO 18640-1:2018 states that the main function of protective clothing for firefighters is to protect the wearer from hazards and to maintain their health and comfort. In addition, by protecting the wearer from heat and flame in extreme environmental conditions, protective clothing prevents health risks or even life-threatening heat stress. Today's standards specify requirements for the protective performance of protective clothing against heat and flame. The higher the protective performance of such garments, the less heat from the human body is dissipated [6].

The energy demand of firefighters in different situations can exceed 500 W/m2 [7]. Of this, 75–85% is emitted in the form of heat, which has to be dissipated from the human body by thermoregulatory processes in order to avoid an increase in core body temperature [8]. If heat dissipation is not limited, the human body can maintain a normal temperature of between 36.5 and 37.5 °C [9]. Under harsh environmental conditions and/or with limited heat dissipation due to protective clothing, the human body is unable to maintain a normal core body temperature and suffers from heat stress. The ability to work is progressively reduced and further increases in core body temperature can become life-threatening [10]. In order to reduce the risk of heat stress during high-intensity physical activities, the effect of protective clothing on human thermoregulation and the occurrence of heat stress should be assessed [11].

Occupational health and safety at the Rescue Board is governed by safety regulations, the requirements of which must be complied with by the employees in their daily work as well as during drills and training. Familiarisation with the safety manuals is carried out as part of a specific special competence module or training course and is subsequently available to all staff. Additional training and drills are organised to maintain these standards. In addition, ensuring the retention of skills is managed internally by scheduling different training exercises during working hours.

2.2 The Role of Ergonomic Movements in Ensuring Safety at Work

Clothing measurements can become a critical determinant of safety and performance in extreme conditions. For example, workwear used in industrial applications that has a very loose pattern can become entangled in machinery and cause accidents. Garments that are too tight for the wearer are also a problem. Anders et al. 2005 showed that trousers that are too tight or smaller than the wearer's actual size can restrict hip movement and alter the body's muscular activity both at work and during leisure time [12]. Yoo and Yoo 2012 concluded that any clothing-induced restrictions in the pelvic and hip area can cause movement and biomechanical changes in the rest of the joints that are not specifically caused by the clothing item [13]. Going further into the topic, the 2013 study by Eungpinichpong et al. refers to the fact that the habitual wearing of tight trousers by young workers when performing specific job tasks leads to restricted and altered movement patterns in the hip and pelvic region, respectively. These, in turn, may be the cause of increased complaints of lower back pain and work incapacity among young workers. Other problems reportedly caused by wearing very tight trousers include neurological and digestive problems, as well as bladder problems in both men and women [14].

Given the tasks of firefighters and the nature of the situations they work in, firefighters should wear clothing that is more conducive to movement when working. The wearer should remain focused on the incident, but should not be restricted or hindered by the clothing. Clothing that is too tight and/or of a smaller size may expose different areas of the body, causing discomfort to the wearer and increasing the amount of heat reaching the skin, for example. Exposure of body areas is also likely to increase the risk of catching colds, due to being more affected than usual by environmental factors. Similarly, clothing that is too small prevents the wearer from performing the necessary tasks and positions, and increases the likelihood of dangerous situations occurring. Clothing that is larger and/or longer than the normal size can be a hindrance, for example by getting caught up in objects in the outdoor environment. In order to design garments that truly fit the body they are intended for, a thorough understanding of the body shapes and sizes of the target audience is necessary. The first step in this process is the systematic body measurement of a typical sample of the population also known as anthropometry. Given that anthropometry can be used to study and analyse the size and shape of the human body, this knowledge can be used to design products that meet the users' needs. Clothing must meet the wearer's requirements for appearance and comfort, allowing them to perform their tasks without constraints or obstacles. Finding the right fit for all users is of course complicated due to the fact that people differ not only in terms of their body dimensions, shapes and proportions, but also in their perceptions of good fit and appearance of clothing [15].

3 Development of an Anthropometric Method for Firefighters' Uniforms

3.1 Research Methodology and Sample

The study involved semi-structured and structured interviews and purposive sampling to collect data. Assessments were carried out in different fire brigades according to predefined questions and a 3-point scoring system. The scale interpretations of the responses are as follows: "1"—no change; "2"—little change; "3"—significant change.

The assessor pointed out that a little change for a score of "2" is visible and/or noticeable to the firefighter, but a visible change for a score of "3" is already visible to the assessor.

Interviews were carried out individually in the teams. Fourteen shifts from four rescue brigades took part in the evaluation and 61 firefighters evaluated the station uniform. The interviews conducted in the rescue brigades took on average 3–4 h, regardless of the size of the brigade. For the time of the visit to the brigades, a figure with 12 body positions had been created, for each position an evaluation sheet with the parameters to be assessed and relevant questions had been prepared using the MS Excel 2016 program. The questions for each parameter were structured as simply as possible so that they could be clearly understood by the firefighters. For example, for the sleeves parameter, the question "Did the sleeves rise up?", for the deltoid parameter the question "Does it feel tight in the deltoid area?", for the abdomen parameter the question "Was the abdomen exposed?" was asked.

During the interview, the firefighters were able to talk about their experiences and the problems they encountered while wearing the clothes. They were open with the assessor and spoke about their observations on the characteristics of the station uniform based on their experiences, which gave new ideas to solve problem areas.

3.2 Characteristics of the Firefighters' Station Uniforms

The station uniform worn by firefighters on a daily basis consists of five different items of clothing: T-shirt, polo shirt, blouse, trousers and sweatpants. The polo shirt is not part of every firefighter's kit, but is worn by team leaders.

The work of firefighters is very varied and they have to work in very tough conditions. Their uniforms are the same all year round, consisting of a station uniform and firefighting overalls. It should also be noted that, wearing the same clothing, firefighters must be prepared to work in temperatures ranging from +25 degrees Celsius to −25 degrees Celsius. Extreme winter colds are less frequent, but it is important to be aware of the temperature range and to bear in mind that, at the same time, firefighters are performing their physical work in full equipment: undergarments, station uniform, firefighting overalls, boots, socks, gloves, balaclava and helmet. In addition, depending on the incident, they carry a variety of work equipment, such as fire hoses, breathing apparatus, hydraulic tools, etc., with them.

This research focuses in particular on the wearing comfort of the T-shirt and trousers. The men's T-shirt has a straight silhouette, short sleeves and a bottom hem that is curved and elongated in the back of about 35–40 mm (Fig. 4). The women's T-shirt has a semi-fitted silhouette with short sleeves and a straight bottom hem (Fig. 5). There is a small monochrome grey print on the left side of the front chest line and a large monochrome grey print on the upper back. On the left sleeve, the

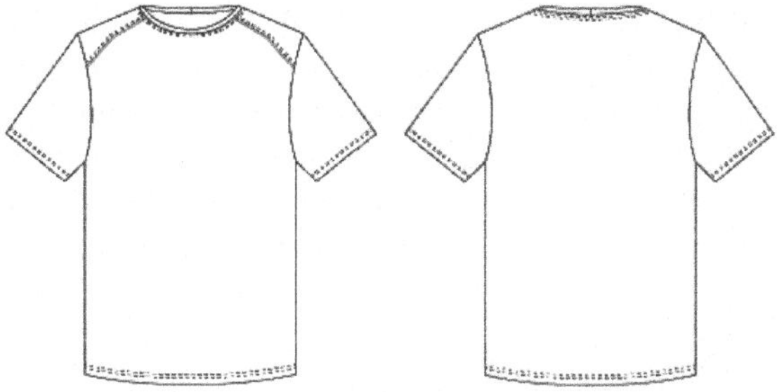

Fig. 4 Men's T-shirt front and back view

Fig. 5 Women's T-shirt

logo of the Rescue Board is printed in white as an outline, with a diameter of 67 mm. The material is 62% polyester and 38% cotton.

Men's uniform trousers have a straight silhouette, a front fastening with a zip, a straight waistband and eight pockets. The front fastening is a 6 mm spiral zipper. The side pockets are diagonally angled, i.e. slanted, and there are two rear pockets with flaps at the back, thigh pockets with sewn-on flaps above the knee and a mobile pocket with bellows and a flap sewn onto the right thigh pocket. The left thigh pocket has a two-pencil-wide stitched pocket and a textile-zip stripe between the side seams to adjust the width of the trouser leg.

Women's trousers have a straight silhouette, a front zip fastening, a figural waistband and four pockets. The front fastening is on the men's side and also closes with a 6 mm spiral zipper. The side pockets are slanted and there are two rear pockets with flaps on the back. Both men's and women's trousers are made of the same material, with a fibre composition of 60% modacrylic and 40% cotton.

3.3 Anthropometric Method for the Firefighters' Station Uniforms

For the assessment guide for the uniforms, body positions were determined according to the content of the firefighters' work tasks. Different activity scenarios were considered and the most important positions were selected. The structure of the guide was based on the international SWW (Smart and Safe Work Wear) project, which had outlined important observations to be observed in uniforms [16]. Among other things, the SWW project resulted in the development of a guide for assessing the fit of uniforms, which, in addition to measurement charts and reminders for assessing the characteristics of the garment, also included drawings that the individual can follow to assess the fit. In addition, inserted between the drawings were the necessary observations to be taken into account by the wearer when assessing the suitability of the uniform.

The model wore a T-shirt and trousers from the set when the body positions guide for firefighters was prepared. 12 body positions were included in the guide developed during the research: sitting, bending over, hands at the side, hands up, hands in front, hand over shoulder, full squat, half squat, crooked kneeling, kneeling on one knee, lunge, crawling (Fig. 6).

The first position was the sitting position, which applies to all firefighters, regardless of their specific abilities or the nature of the incident. The assessed parameters were sleeves, lower back, crotch and trouser cuffs. The contraction of the sleeves shall be monitored in order to detect the presence of possible tension on the upper back. In the crotch area, both the crotch seam area and the thigh area are assessed. When sitting, the point of tension is the buttock area and the tension manifestation point is the lower back. In addition, the contraction of the trouser legs shall be monitored to identify any points of tightness in the leg region.

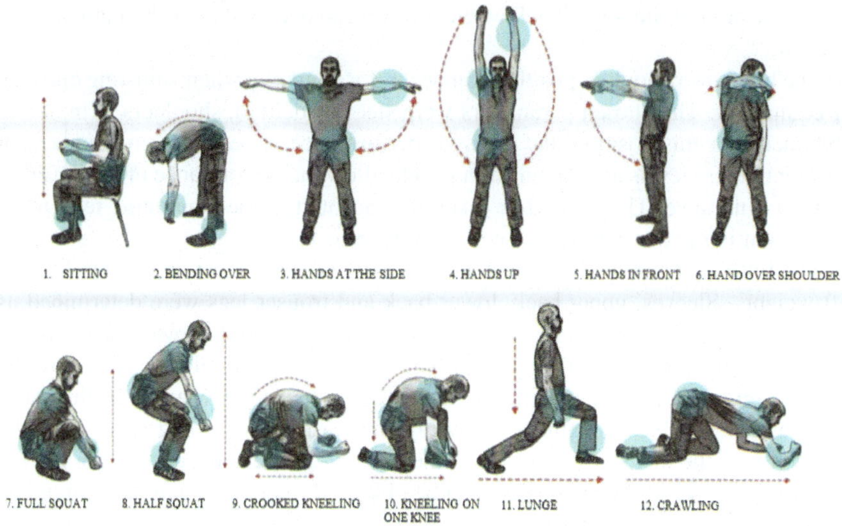

Fig. 6 Guide to body positions

The position of bending over was also added, as it is often encountered in the work of firefighters. Already on the way to a site, firefighters bend over to put on their firefighting overalls. In addition, bending over occurs in the course of many tasks: taking fire hoses from the ground, cutting trees with a chainsaw, handling hydraulic tools at a traffic accident scene, etc. The assessed parameters were sleeves, upper back, lower back and trouser cuffs. It is important to observe the exposure of the wearer's back when assessing, for example, when testing a new item of the station uniform, to provide feedback on the characteristics of the material of the station uniform, its fit in terms of pattern and its parameters. The upper back and sleeves parameter was added to identify the tension on the upper back.

Hands at the side and hands up positions go hand in hand, but are assessed separately on the same parameters. Firefighters often encounter situations where they have to raise their arms, such as when entering the fire engine, opening and taking things from the fire engine's compartment, climbing a ladder, etc. The chosen parameters were sleeves, deltoid, armpit and abdomen. Greater attention is paid to the upper back area, where potential points of tension are monitored. It is important that in these types of positions the garment does not constrain the wearer as these positions are frequently used. It is also important to monitor abdominal exposure in order to identify any potential problem there in terms of material properties or shirt length when testing new station uniforms.

Hands in front position is often common, for example, when taking or handing things to others, climbing a ladder, taking things from the fire engine's compartment, etc. Because of the high frequency of this position, it is an important position in the list of assessed positions. The selected parameters were sleeves, deltoid and upper back. The area of the upper back where tension can accumulate is the main area of interest.

The hand over shoulder position can occur, for example, when adjusting clothing or equipment. The chosen parameters were sleeves, deltoid, shoulders, armpit and abdomen. For this position, the assessed performance varies between people with different fitness levels and the uniform should allow the performance of this position without hindrance. This should be taken into account in the ergonomic tests to be carried out in the development of new station uniforms.

The full squat and the half squat also go hand in hand, most commonly used to lift weights. Sleeves, upper back, lower back and trouser legs were determined as the parameters. The potential strain on the upper back and the lower back and leg area are the most significantly assessed. The specific positions clearly show how the station uniform affects the wearer as their body dimensions change. In this case, for example, the back length of the wearer increases when doing squats and the suitability of the material properties of the T-shirt, blouse and trousers, the pattern of the garments and the parameters can be observed to be taken into account when designing new station uniforms.

The crooked kneeling position is most likely to occur during a rescue with a life safety rope, but similar positions also occur in the event of emergencies. The chosen parameters were sleeves, deltoid, upper back, lower back and trouser cuffs. When conducting ergonomic tests, the properties of the uniform's material should be assessed, taking into account that the garment should allow the wearer to assume a position without hindrance.

The kneeling on one knee position is assumed by almost every firefighter connecting fire hoses. The chosen parameters were sleeves, upper back, lower back and trouser cuffs. Potential points of tension on the upper and lower back are assessed. In ergonomic tests, back exposure will be monitored to verify the suitability of the material properties and pattern solutions of the trousers and T-shirt.

The lunge is most commonly used to describe getting into the fire engine, where a long lunge is performed, simultaneously extending the arms. For the emergency events, the firefighter may need to perform knee lifts and lunges, for example, by stepping on a window sill, climbing a ladder or other surface depending on the incident. The chosen parameters for this position are the crotch and the trouser leg. As this position places high demands on the material of the trousers, the pattern and its various parameters, ergonomic tests should be carried out to monitor the extent to which the garment imposes limits on the wearer.

Crawling is the most common position for smoke diving. While smoke diving is performed differently among the rescue workers, depending on the specifics of the event and the personal preference of the firefighter, crawling on elbows and knees was one of the possible variations performed during the assessment. Sleeves, shoulders, upper back, lower back and trouser cuffs were defined as parameters. The expectation is to assess the station uniforms during the movement in this position in order to identify possible areas of tension in the different parts of the body.

4 Anthropometric Assessment and Analysis of Station Uniforms

For the sitting position, a total of four parameters were assessed: sleeves, lower back, crotch and trouser cuffs (Fig. 7). By observing these parameters, potential tension points can be identified. For example, tension applied to the shoulder blades will shorten the sleeves, tension applied to the buttocks will expose the lower back, etc. For the crotch parameter, both the tension at the crotch seam and the tension at the inner thigh were assessed. Lastly, the lifting of the trouser legs was assessed, which was rated as "3" by 41 people. This parameter was also actively monitored by the assessor, who gave their own assessment of the result by observing the situation. When sitting, the fact that the knees were pulling the fabric was repeatedly mentioned. In some cases, the problem was the inclusion of indentations that should allow the knee to bend, but probably due to the characteristics of the material, this caused discomfort to the user.

In the bending over position, there were a total of four parameters to be assessed: sleeves, upper back, lower back and trouser legs (Fig. 8). For sleeves, 58 people gave a score of "3", which is an overwhelming majority. The next parameter was the upper back, which was rated "2" by 40 people. The assessor considers that the visible shortening of the sleeves also reduced the perceived tension in the upper back. In the bending over position, it was noted that the main tension on the T-shirt is perceptible between the shoulder blades. For the lower back, 29 people gave a score of "2" and 16 people gave a score of "3", noting that they often had to adjust their trousers after sitting. In some cases, the trousers did not drop down visibly, but the back of the trousers began to visibly pull away from the body, exposing the lower back.

The third is the hands at the side position, where four parameters were assessed: sleeves, deltoid, armpit and abdomen (Fig. 9). For this position, it was pointed out that the combination of the T-shirt and blouse caused discomfort. From the assessment results it can be seen that there were very few instances of tension in this position.

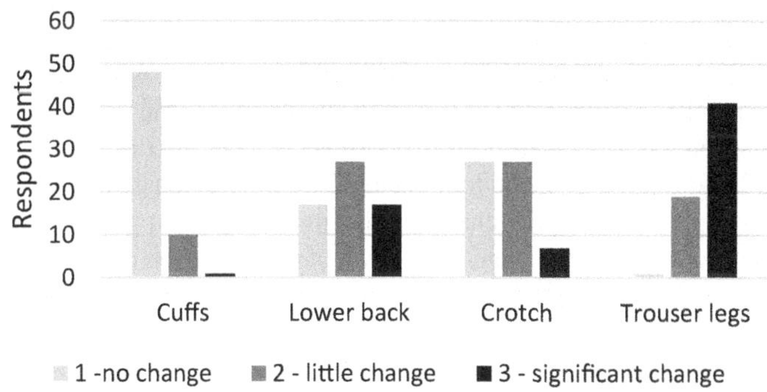

Fig. 7 Sitting

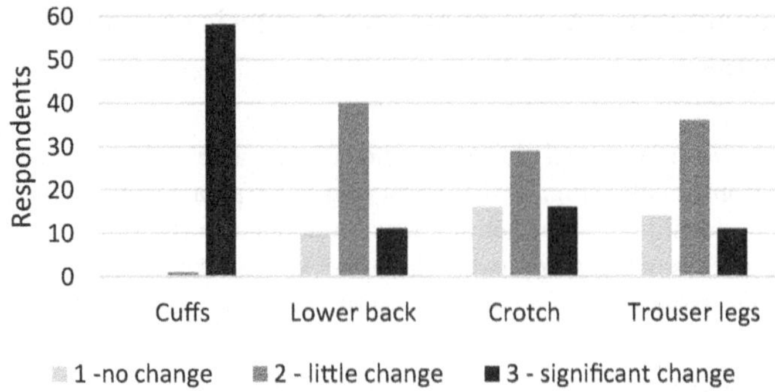

Fig. 8 Bending over

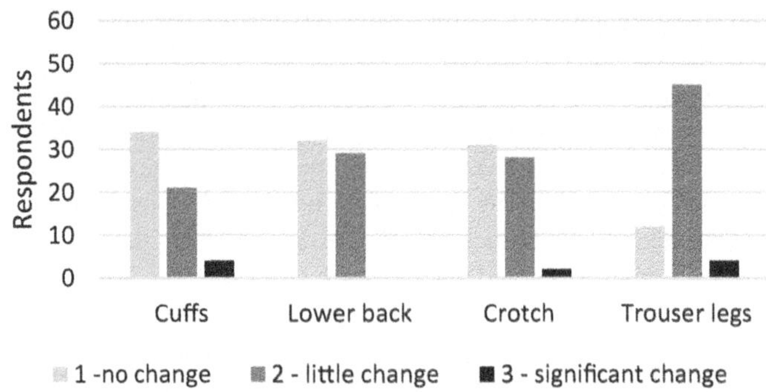

Fig. 9 Hands at the side

The scores of "1" and "2" do not need to be reported separately as in this case the garment did not constrain the wearer.

The next body position after hands at the side was to bring the arms fully up to a vertical position (Fig. 10). In the hands-up position, the same parameters were assessed as in the hands at the side position. Compared to the previous position, the results changed significantly. The most significant change was in the sleeve and abdominal area. Previously, four subjects rated the change in the length of the sleeves as "3", but when the arms were raised, 49 subjects gave a score of "3". Similarly, there was a large change in the abdominal area, where previously a score of "3" was given by 4 people and after the change in position 46 people gave a score of "3". There was also a noticeable change in the level of tension in the deltoid and shoulder region. The firefighters pointed out that the fabric of the T-shirt was bulking up on the shoulders, pulling on the armpits and sides. This position shows an increase in discomfort for the wearer due to the characteristics of the material.

For the hands in front position, three parameters were assessed: sleeves, deltoid and upper back (Fig. 11). The most noticeable changes are in the shortening of the sleeves. For this parameter, 32 people gave a score of "3". The assessor considers

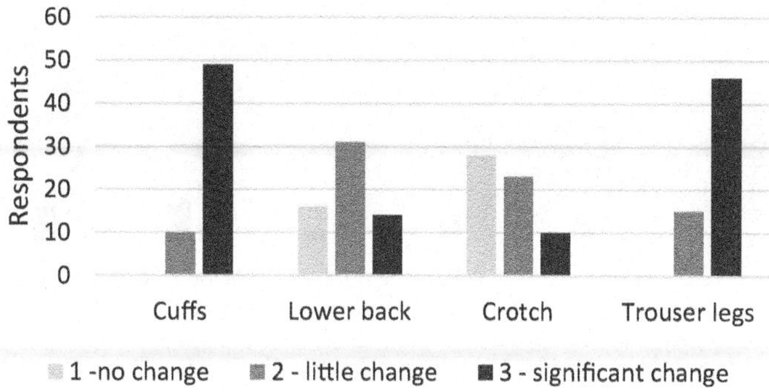

Fig. 10 Hands up

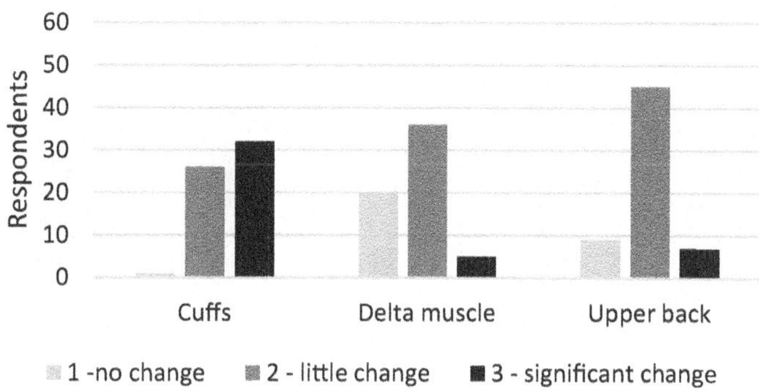

Fig. 11 Hands in front

that the shortening of the sleeves was a direct compensation for the strain on the shoulder blade area. During the assessment, the firefighters also pointed out that in this position there is tension under the armpits and in the deltoid region, which causes discomfort. As the fabric of the T-shirt is not elastic, the tension built up in the area of the seams and caused the user to feel uncomfortable.

For the hand over shoulder position, five parameters were assessed: sleeves, deltoid, shoulders, armpit and abdomen (Fig. 12). When looking at the results, the most significant change can be seen in the armpit and abdomen regions. The firefighters also reported a feeling of tension in the armpit area, a build-up of cloth on the shoulders and tension in the upper back area.

In the full squat position, there were four parameters to be assessed: sleeves, upper back, lower back and trouser legs (Fig. 13). From the diagram, it can be seen that the position in question created quite noticeable points of tension in different areas of the body. The most noticeable areas of tension are the sleeves, lower back and lower legs. For all the above parameters, half of the respondents have given a score of "3", indicating a visible change. A large number of firefighters noted a pulling sensation in the knees and thighs in this position. One firefighter pointed out that

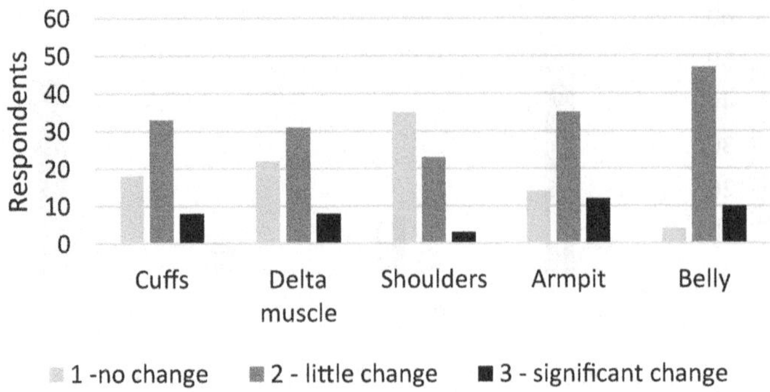

Fig. 12 Hand over shoulder

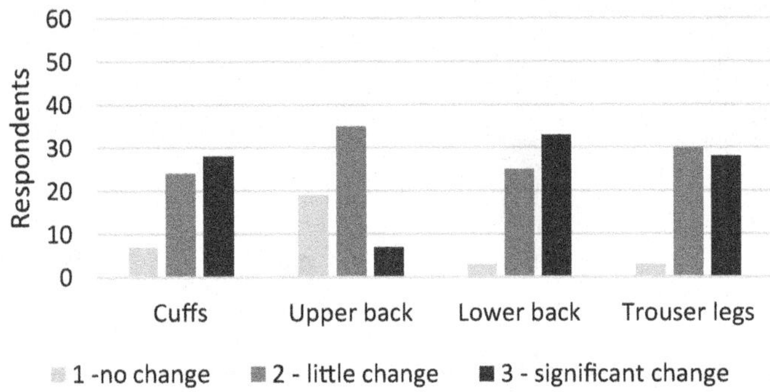

Fig. 13 Full squat

when squatting, the fabric accumulates at the back of the knee and even starts to rub. This was said to cause discomfort, for example, on the back of the jet ski during prolonged squatting. This position is more difficult and places higher expectations on the garment. There is a high level of tension on the upper and lower back, which is reflected in the values of the observable parameters. For example, to compensate for the strain on the upper back, the sleeves are shortened in many cases. Due to the lack of elongation in the fabric, the garment has become a hindrance to the wearer when performing physical activities.

In the half squat position, the garment was assessed on the same parameters as in the full squat position (Fig. 14). The bar chart shows a slight change in the feeling of tightness by the respondents. A decrease in the shortening of the trouser legs is most noticeable and the tension on the lower back has also slightly decreased.

In the crooked kneeling position, the uniform was assessed on five parameters: sleeves, deltoid, upper back, lower back and trouser legs (Fig. 15). The graph shows how the maximum values are of high proportion. The most noticeable changes are in the sleeves, upper back and lower back. Tension under the armpit was mentioned several times in the feedback. Similarly, two firefighters pointed out that it is difficult

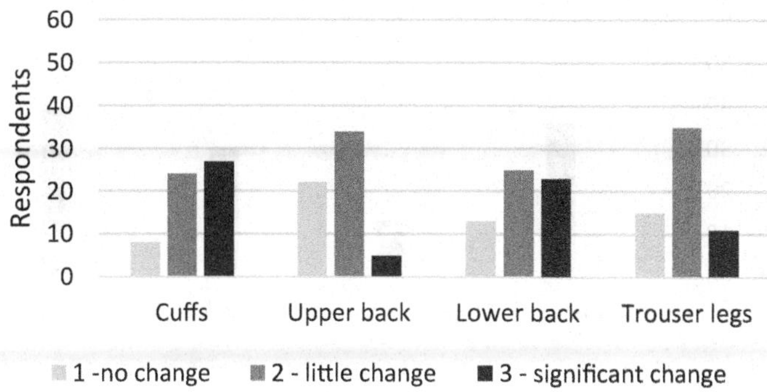

Fig. 14 Half squat

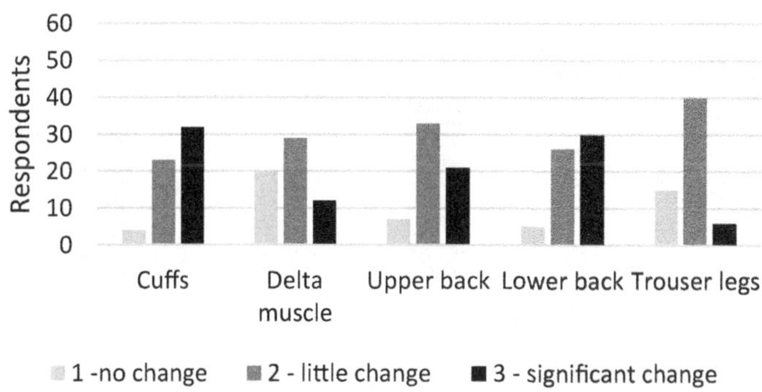

Fig. 15 Crooked kneeling

to be comfortable in the uniform in an emergency situation where the victim has to be placed in a steady sideways position on a "log-roll" principle. In the position in question, the dimensions of the body areas have changed significantly and the lack of elasticity of the fabric has led to the garments becoming a hindrance to the user performing the task.

For the kneeling on one knee position, four parameters were assessed: sleeves, upper back, lower back and trouser legs (Fig. 16). The most significant changes were seen/felt with the sleeves, lower back and trouser legs. For the above parameters, at least half of the respondents gave maximum scores. The firefighters reported feeling a pulling sensation in the thighs and buttocks. In addition, the T-shirt felt as if it was pulling and restricting the wearer's movement. Once again, the characteristics of the fabric play a role in the results, causing discomfort to the wearer and restricting their freedom of movement.

For the "lunge", two parameters were assessed: crotch and trouser legs (Fig. 17). In the case of the lunge, the assessor noticed how the firefighters automatically raised their trousers higher to assume the given posture. The firefighters were then asked to drop their trousers back to the original position and the assessment was started

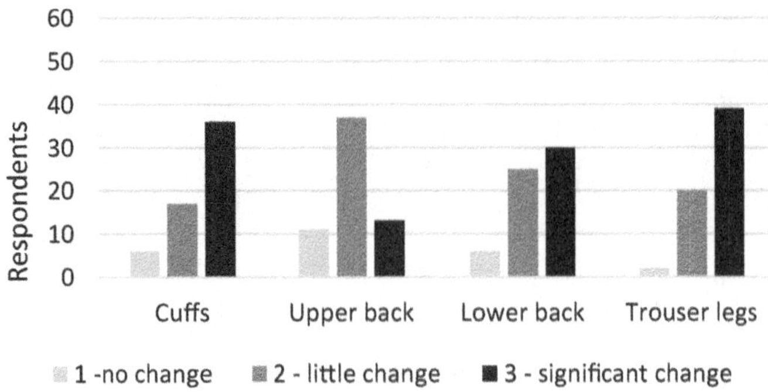

Fig. 16 Kneeling on one knee

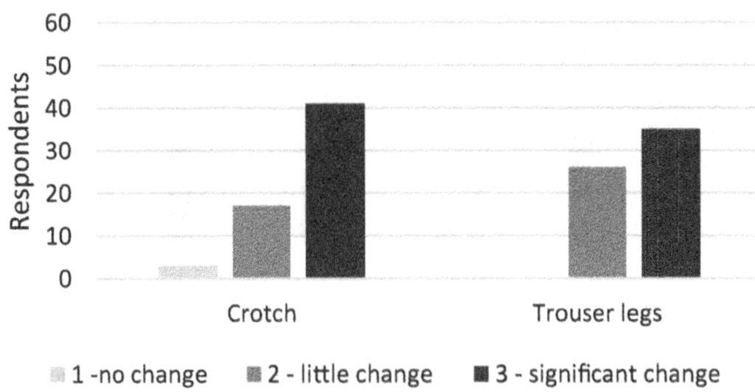

Fig. 17 Lunge

again. Of all the firefighters who assessed the uniform, 41 gave a maximum score of "3" for crotch area tension. The firefighters pointed out that the fabric in the crotch area was at its limit and that there was also a noticeable pulling sensation in the thigh area. They pointed out that when the trousers are wet, the elongation of the fabric is even worse. Several workers in the test mentioned having to push through the resistance created by the trousers in order to achieve a given posture. Looking at the results, it can be said that the trousers restrict the wearer's freedom of movement due to the poor elongation of the fabric. In addition, the pockets and the various seams contribute to increasing the resistance to movement.

For the crawling position, five parameters were assessed: sleeves, shoulders, upper back, lower back and trouser legs (Fig. 18). For this position, the firefighters made several observations: thigh area is tight, incisions on the knees are rubbing, fabric is pulling away from the knees, buttocks area is tight, pulling under the armpit. In the case of the parameters for the sleeves, lower back and trouser legs, half of the respondents gave a maximum score of "3" for the change. The upper back also scored higher and it can be seen that in this position, several areas of the body become very tense and create resistance to movement for the wearer.

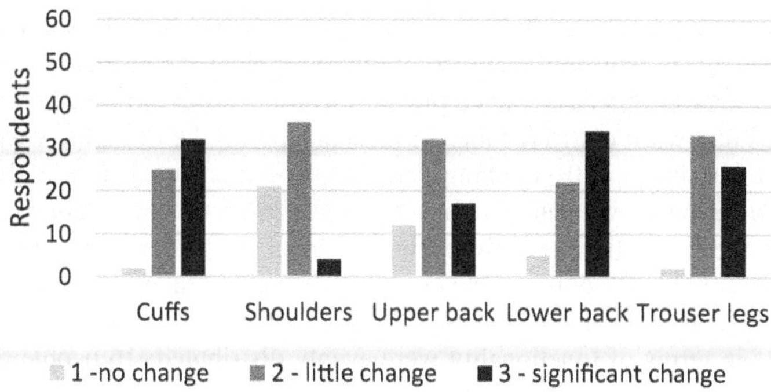

Fig. 18 Crawling

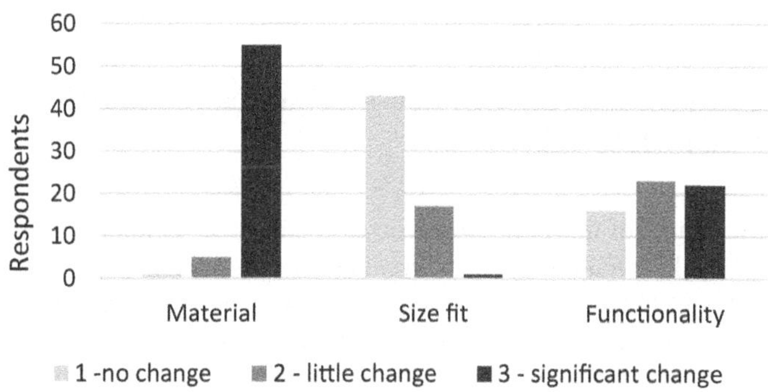

Fig. 19 Other factors

In addition to the dynamic anthropometry method, the assessment included the collection of data on other factors, the analysis of which will help to assess the suitability of the uniform for the work environment (Fig. 19).

It was pointed out that clothing materials make workers sweat when they perform their tasks. In cold weather, this has a major impact as the wearer quickly becomes cold when the garments against the body are wet. Problems with the fit of the garment were brought out. As the overgarments of rescuers consist of firefighting overalls and accessories, they cannot adjust the uniforms under the overalls during work. The assessor noticed that many workers lift their trousers before sitting down. Investigating the matter, it emerged that, without having lifted the trousers beforehand, the garment starts to pull at the knees and lower back due to poor elongation. As many as 55 respondents said that a change in materials was necessary. Other factors also included the suitability of the size of the firefighters' clothing. Looking at the results, it can be observed that the sizing is mostly correct. The main items of clothing that were the wrong size were T-shirts that were too tight around the body. It was also noticed that trousers were taken a number larger in many cases to deliberately create a loose fit. In such cases, a belt was used to prevent the trousers from falling down. The author estimates that 43 of the 61 participants had chosen uniforms of an appropriate size. For only one worker did the evaluator consider it important to change the size. In this case, the employee's trousers were too long and the lower trouser legs had worn out over time. At the same point, the firefighters were asked about the functionality of the clothing. The interviewees stated that the mobile phone pocket with bellows sewn onto the trousers' thigh pocket should be larger, as modern phones are too big to fit into it. Regarding the inner pockets, it was noted that they are not very suitable for putting objects in, as the space is taken up by leg room and therefore prevents the performance of various movements (e.g. bending, squatting). The results in terms of functionality were evenly distributed: 16 people rated it as "1", 23 as "2" and 22 as "3". It can be concluded that the need for change is not as critical as the need for a change in material, but that these observations should be taken into account when designing new station uniforms.

5 Determination of the Physico-mechanical Properties of T-Shirt Materials

Based on the results of the anthropometric assessment of station uniforms, the firefighters identified the material of the current uniforms as the main problem. The vast majority of the firefighters stated that the uniforms make them sweat very easily and that the material does not transfer moisture away from the body well enough, leaving the user feeling wet. It was also noted that when wet, the garments do not elongate enough and are uncomfortable to perform tasks. For example, when the hand is lifted, the material builds up on the shoulder and when the hand is lowered again, the fabric of the T-shirt does not return to its original position.

As the Rescue Board is in the process of developing uniforms, it was decided to start by investigating the properties of new materials for the T-shirt. Tests were also carried out on the existing T-shirt fabric for comparison. Tests were conducted in the Textile Testing Laboratory of the TTK University of Applied Sciences to evaluate the breathability and elongation of the material of the T-shirt of the Rescue Board's uniform.

The tests were carried out on the currently used T-shirt material art TFA 1 and on possible new fabrics. Four new possible options were selected and submitted for testing by the Rescue Board, and all of them were tested against the given parameters of the existing material. The technical specifications of the five different T-shirt fabrics are shown in Table 1.

The air permeability test was carried out according to ISO 9237:2000 [17]. For the determination of air permeability the Textest company's FX 3300 LabAir IV was used. The room temperature at the time of testing was 23 °C and the humidity 32%. These conditions do not meet the requirements of the standard, but since all fabrics were tested under the same conditions, the results are comparable. To find

Table 1 Data of the fabrics

No.	Item	Weight, g/m^2	Composition	Weave construction
1	TFA 1	210	38CO 62PES	Pique
2	TFA 2	190	58MA 40CO 2AT	Single jersey
3	TFA 3	260	58MA 40CO 2AT	Pique
4	TFA 4	160	75CO 25TAC	2 thread fleece
5	TFA 5	180	95CO 5EL	2 thread fleece

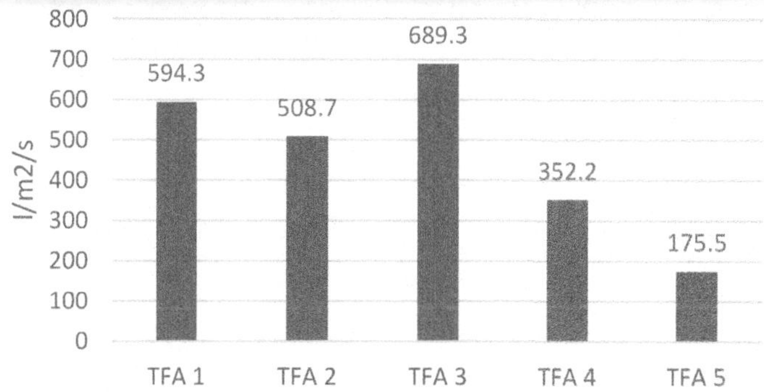

Fig. 20 Air permeability of the five fabrics in comparison

the average value of air permeability, the measurement was repeated 10 times and the air permeability of each fabric was calculated.

Figure 20 shows that, of the four materials selected, only fabric TFA 3 has better air permeability compared to the fabric of the current t-shirt (TFA 1). The air permeability of fabric TFA 2 is about 15% lower than that of the currently used product. The other two materials selected, TFA 4 and TFA 5, have several times lower air permeability. Out of the five materials tested, TFA 5 has more than three times lower air permeability than the current TFA 1 fabric.

In addition to air permeability testing, the elongation of the selected fabrics was also determined. The SDL Atlas Fryma Fabric Extensometer device was used to measure the elongation (Fig. 21) as the aim was not to determine the elongation at break. This test was carried out according to the manual of the device, as there is no ISO standard.

The laboratory conditions were the same as for the air permeability test. As the tests were carried out under the same conditions, the results for all fabrics are comparable.

All the materials tested were knitted fabrics, therefore the test pieces were cut to 75 mm × 90 mm, with an area between the grips of 75 mm. Two test pieces were cut from each fabric, one test piece in each direction (wale and course). A weight of 3 kg was applied to the test pieces.

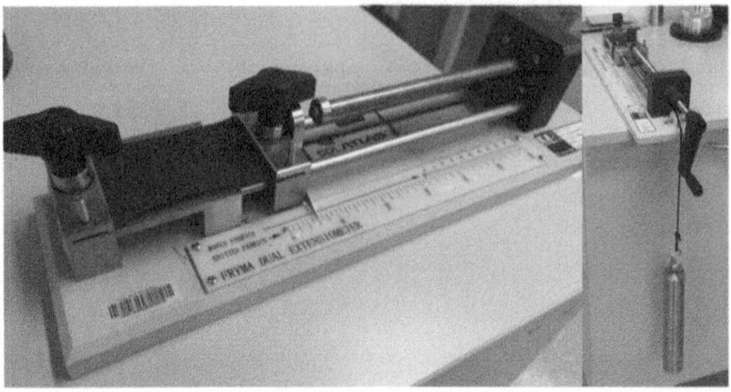

Fig. 21 SDL Atlas Fryma fabric extensometer

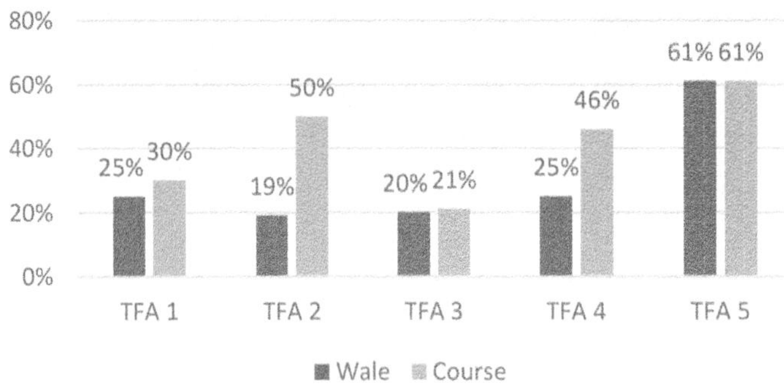

Fig. 22 Elongation in a comparison of five fabrics

Figure 22 shows that the fabric with the best elongation is TFA 5, a cotton/elastane blend with 61% elongation in both directions. The fabric with the lowest elongation is TFA 3, with an elongation of ~20% in both directions. Comparing the material currently in use with the other fabrics tested, three fabrics have the highest elongation. The elongation of the TFA 2 fabric in the course direction is 20% higher than the elongation of the TFA 1 material used, while the elongation in the wale direction is 6% lower. TFA 4 has the same elongation in the wale direction as the material currently in use and 16% more elongation in the course direction.

6 Assessment and Analysis of Ergonomic Anthropometry of the T-Shirt

In the ergonomic tests, the test wearers performed 10 positions (Fig. 6). In the T-shirt tests, the sitting (1) and standing (9) positions were excluded, as they provide the most information on the material properties of the trousers and the suitability of the patterns. The other 10 body positions that were tested occur in the daily work tasks of firefighters. The assessment was performed in a 3-point system with values of "1"—no change, "2"—slight change and "3"—visible change. Testing was based on scoring as follows: a slight change to a score of "2" is visible and/or perceptible to the firefighter, but a visible change to a score of "3" is already perceived by the bystander.

The T-shirt was rated for comfort based on five different areas of the body, which the assessor observed and asked subjects to comment on. The points to be assessed were upper back, deltoid, armpit, lower back and abdomen. Based on the analysis of the research data, Figs. 23, 24, and 25 show the analysis of the T-shirts made from the three new materials selected by the Rescue Board. The evaluations were carried out at the Rescue Board's stations in three different areas of Tallinn. Four test wearers took part in the ergonomic tests and trial wear, all of whom tested T-shirts made of three different materials.

Comparing the ten different ergonomic movement ratings in a 3-point system, the data of which are shown on the vertical axis, we see that the T-shirt made of fabric art TFA 3 shows the most striking changes (score 3) in the different body regions. The least striking changes occur when testing the T-shirt made of fabric type TFA 4.

According to the test wearers, the T-shirt made of art TFA 4 was most similar to T-shirts made of art TFA 1, which had been used in the past. In the ergonomic tests, the test wearers noted that despite the tension in the armpit, due to the structure and composition of the TFA 4 fabric, the good elongation properties of the fabric allowed to reduce the feeling of tension and the discomfort was low. Test wearers' comments on the T-shirt made from this material were as follows: in the crooked kneeling position, the sleeves start to feel tight if left in this position for a long time; there is more tension on the shoulders when in the kneeling on one knee position; the fabric becomes electrified; there is resistance when bending over, but as the shirt is elastic, it stretches; the pattern could be looser.

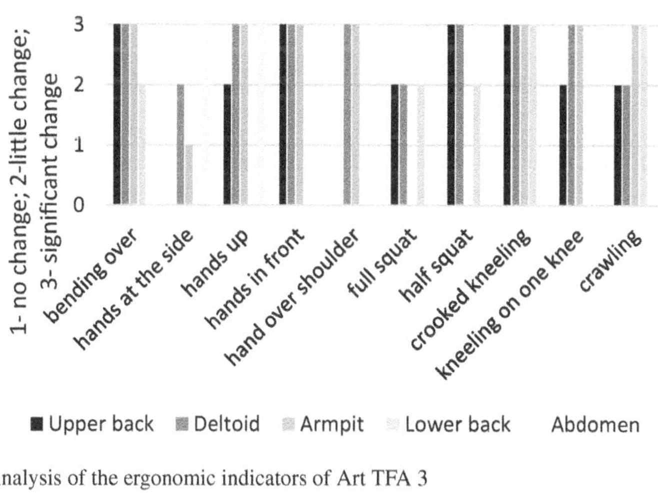

Fig. 23 Analysis of the ergonomic indicators of Art TFA 3

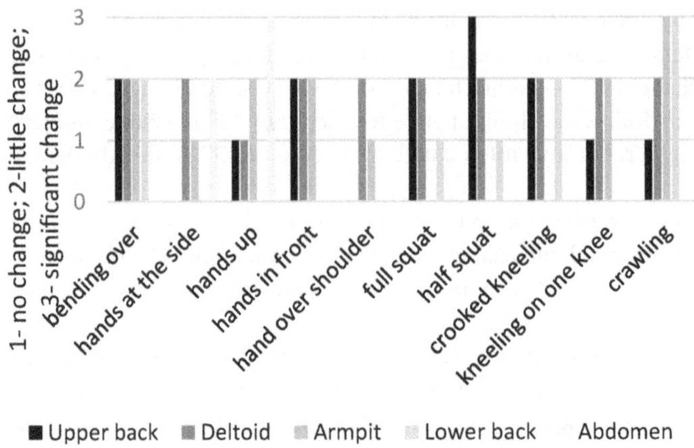

Fig. 24 Analysis of the ergonomic indicators of art TFA 4

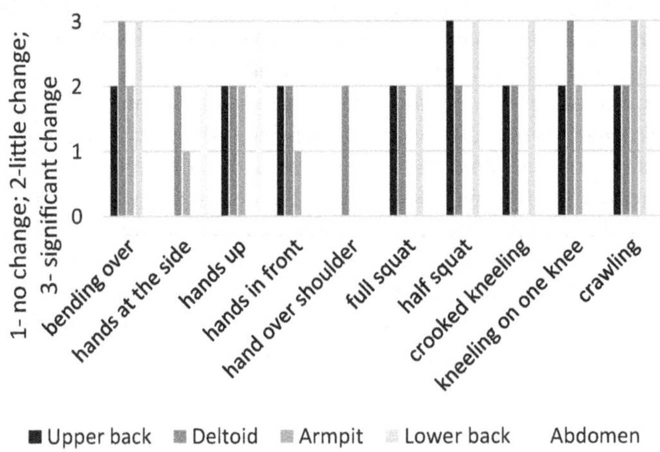

Fig. 25 Analysis of the ergonomic indicators of Art TFA 5

During the test wear, tests were carried out on a rowing ergometer, with all four test subjects performing 30 pulls. This test assessed the change in weight of the T-shirt as a result of the wearers sweating. The following table shows the results of the three fabrics tested, comparing the weight change with the dry shirt weight (Table 2).

Table 2 shows that the T-shirt material TFA 3 changed the least in weight during the physical exercise test. When compared to the air permeability of the material measured in the laboratory, this fabric had the highest air permeability (Fig. 20). According to the test wearers, the T-shirt made of fabric type TFA 4 was the most comfortable to wear. Based on Table 2, the weight of the T-shirt has become almost 2 times heavier during physical exercise compared to material TFA 3. When comparing

Table 2 Weight change of T-shirts in the rowing ergometer test

Article	Composition	Difference in weight with dry shirt weight, g				
		Test 1	Test 2	Test 3	Test 3	Average
TFA 3	58MA 40CO 2AT (pique)	+2	+2	+6	+16	+6.5
TFA 4	75CO 25TAC	+10	+6	+14	+16	+11.5
TFA 5	95CO 5EA	+8	+14	+22	+10	+13.5

the air permeability tests of the materials, TFA 4 has almost 50% less air permeability than fabric TFA 3.

From a physiological point of view, the test wearers' comments on the TFA 4 T-shirt fabric were as follows: the fabric feels the same when sweating as it does when dry, it feels warm on the back rather than cold, the smell of sweat is noticeable, the fabric feels breathable, dries well, comfortable and suitable for workwear.

The breathability of the materials, the elongation when wet and the drying speed of the drenched shirt on the back after physical testing on a rowing ergometer also need to be further investigated.

7 Conclusions

Several problems with the uniforms for firefighters were related to the properties of their materials (air permeability, breathability, elongation, drying speed).

The analysis shows that products with a predominance of natural fibres in their material composition provide comfort. The elasticity and elongation of the material ensures greater freedom of movement for the wearer and does not hinder ergonomics when performing work tasks. When designing a new product, attention must be paid to the necessity and practicality of every detail and seam. It is important to test the effect of both the pattern selection and the material properties on the wearer in ergonomic tests. Considering that firefighters wear station uniforms throughout the day, both in the field and in harsh conditions, the body-friendliness of the materials must be taken into account. It is important to find a suitable fabric with a structure that can help to dissipate heat from the body without creating the potential risk of heat stress, which can reduce the firefighters' performance and in some situations be life-threatening.

To assess the physiological tests, it would be necessary to determine the breathability of the materials. Even though the work of firefighters involves exposure to intense heat, the breathability of the materials may not ensure complete physiological comfort for the wearer. It would be necessary to test the drying rate of the T-shirt when it is drenched in sweat by the wearer due to physical exertion.

During the positions carried out during the interview, it was evident that the fabric of the uniforms is not elongating and prevents the wearer from performing more difficult exercises. The parameters set for the areas of the body that were observed were often at the maximum score for the more difficult positions, where the wearer felt discomfort. During the course of the tasks, the dimensions of the firefighter's body regions change significantly and their clothing should be able to support this process. Based on the results of the research, it is recommended to use a composite fabric with a high proportion of natural fibres for products that rest on the shoulder. Chemical fibres and appropriate material treatments to improve the fabric's properties (durability, dimensional stability, elasticity, moisture dissipation) should be added. In the case of fabrics for station uniforms, the selected fabric must meet the requirements for working at high temperatures.

Based on this research, it is recommended that thorough ergonomic tests are carried out on each product using the guide developed during the research. It is recommended to include 2 additional positions in the guide. For a product that rests on the shoulder, an ergonomic method should be added to the test from the "arms in front" to the "arms up" position. In the case of trousers, it is recommended to include a "knee lift" position. During the ergonomic tests, the parameters of the body regions to be assessed should be added to obtain a more accurate picture of possible problem areas. Parameters assessed to a maximum score of "3" should be critically reviewed in the light of the pattern design and the properties of the fabric.

References

1. Gupta D, Zakaria N (2014) Anthropometry, apparel sizing and design. Woodhead Publishing Limited, Cambridge
2. Bye E, Labat KL, Delong MR (2006) Analysis of body measurement systems for apparel. Cloth Text Res J 24(2):66–79
3. EVS. Basic human body measurements for technological design [Online]. Available: https://www.evs.ee/et/evs-en-iso-7250-1-2017
4. Coca A, Williams WJ, Roberge RJ, Powell JB (2010) Effects of fire fighter protective ensembles on mobility and performance. Appl Ergon 41(4):636–641
5. Boorad ML, Barker J, Lee Y-A, Lin S-H, Cho E, Ashdon PS (2013) Exploration of firefighter turnout gear. Part 1: Identifying male firefighter use needs. J Text Apparel 8(1):1–13
6. Holmer I, Kuklan K, Gao C (2006) Test of firefighter's turnout gear in hot and humid air exposure. Int J Occup Saf Ergon 12(3):279–305
7. Holméra I, Gavhedb D (2007) Classification of metabolic and respiratory demands in fire fighting activity with extreme workloads. Appl Ergon 38(1):45–52
8. Gaesser GA, Brooks GA (1975) Muscular efficiency during steady rate exercise: effects of speed and work rate. J Appl Physiol 38(6):1132–1139
9. Mahmood MA, Zweifler M (2007) Progress in shivering control. J Neurol Sci 261:47–54
10. Koelblen B, Psikuta A, Bogdan A, Annaheim S, Rossi RM (2017) Comparison of fabric skins for the simulation of sweating on thermal manikins. Int J Biometeorol 61(9):1519–1529
11. EVS. Protective clothing for firefighters—physiological impact—Part 1: Measurement of coupled heat and moisture transfer with the sweating torso [Online]. Available: https://www.evs.ee/et/evs-en-iso-18640-1-2018
12. Anders C, Scholle HC, Wagner H, Puta C, Grassme R, Petrovich A (2005) Trunk muscle co-ordination during gait: relationship between muscle function and acute low back pain. Pathophysiology 12(4):243–247
13. Yoo IG, Yoo WG (2012) Effects of wearing of tight jeans of lumbar and hip movement during trunk flexion. J Phys Ther Sci 24(8):659–661

14. Eungpinichpong W, Buttagat V, Areeudomwong P, Pramodhyakul N, Swangnetr M, Kaber D, Puntumetakulae R (2013) Effects of restrictive clothing on lumbar range of motion and trunk muscle activity in young adult worker manual material handling. Appl Ergon 44(6):1024–1032
15. Pheasant ST, Haslegrave CM (2006) Anthropometry, ergonomics and the design of work. Taylor and Francis, CRC Press, London
16. SWW—Smart and Safe Work Wear. Centria [Online]. Available: https://tki.centria.fi/hanke/sww-smart-and-safe-work-wear/7710/7710/7710
17. EVS. Textiles—Determination of the permeability of fabrics to air [Online]. Available: https://www.evs.ee/et/evs-en-iso-9237-2000

Development of Modified Polymethyl Methacrylate and Hydroxyapatite (PMMA/HA) Biomaterial Composite for Orthopaedic Products

Umang Dubey, Shivi Kesarwani, Panagiotis Kyratsis, and Rajesh Kumar Verma

PMMA bone cement is an essential part of orthopedic component fabrication due to its magnificent biocompatibility and mechanical properties. It is extensively used in the biomedical sector as synthetic bone material. This chapter explores the feasibility of PMMA by the supplement of Hydroxyapatite (HA) based bone cement at varying ranges of 4–16 wt.%. The supplement of 10 wt.% HA nanoparticles to the powder of PMMA bone cement increased flexural strength by 12.8% and compression strength by 21.2%. The distribution of reinforced nanoparticles in the nanocomposite was studied using Fourier transform infrared (FTIR) spectroscopy and X-ray diffraction (XRD). With the help of scanning electron microscopy (SEM) study of the fractured surface of the proposed nanocomposite bone cement, the reason behind the change in mechanical properties was observed. The findings of mechanical testing and advanced characterization demonstrate the application potential for developing orthopedic implants in the biomedical sector.

1 Introduction

Polymethyl methacrylate (PMMA) is widely used in biomedical functions for removable partial and complete prostheses implants [1]. Because of its aesthetic and physical properties, PMMA was widely preferred. Compared to other polymers, PMMA

U. Dubey · S. Kesarwani · R. K. Verma (✉)
Materials and Morphology Laboratory, Department of Mechanical Engineering, Madan Mohan Malaviya University of Technology, Gorakhpur 273010, India
e-mail: rajeshverma.nit@gmail.com

P. Kyratsis
Department of Product and Systems Design Engineering, University of Western Macedonia, 50100 Kila Kozani, Greece
e-mail: pkyratsis@uowm.gr

P. Kyratsis et al. (eds.), *Advances in Product Design Engineering*, Management and Industrial Engineering,
https://doi.org/10.1007/978-3-030-98124-2_7

resins exhibit good flexural strengths and impact strengths [2]. PMMA bone cement was the primary agent of fastening joint prostheses in the human body for more than 50 years.

1.1 History and Development of Acrylic Bone Cement

In 1902, Dr. Otto Rohm developed PMMA bone cement which looks like a glass-like solid substance with excellent biocompatibility. After that, the study of Kulzer in the year 1936 demonstrated that combining a liquid monomer and PMMA powder created a mixture in dough form. It was efficiently cured with the supplement of benzoyl peroxide (BPO), and the mixture was heated to 100 °C. PMMA has been utilized since 1930 for the study of the structure and diseases of teeth and bone cement for implant fixation [3]. In 1938, PMMA was used to treat cranial abnormalities in monkeys for the first time. As surgeons gained a better knowledge of the PMMA system, they began to use it in human reconstructive surgery. In 1949, for the first time, PMMA was used to create a hip prosthesis by the Judet brothers [4]. But it wasn't long before it became clear that due to biological and mechanical constraints, the PMMA prosthesis they were using couldn't be adequately articulated or integrated into the body. Sir John Charnley used self-polymerizing PMMA bone cement to effectively secure an intramedullary stem prosthetic in 1958 [5]. The substance was given the term acrylic-based bone cement by Charnley. Charnley was the first to disclose a revolutionary surgical method for complete hip replacement surgery in 1970. In the early 1970s, antibiotics were added up to bone cement to minimize periprosthetic infection, which is possibly the most terrible complication after total joint replacement. Since Charnley developed PMMA bone cement, more than a million people's lives have been drastically and amazingly transformed due to his invention.

PMMA bone cement is generally used to hold joint prostheses in place. It covers the gap between the bone and the prosthesis during joint replacement fixation and serves as an essential interface (Fig. 1). Bone cement will equally absorb the stresses applied to the bone due to its optimal elasticity. Because of the tight relationship between the cement and the bone, the stresses and interface strain energy are distributed optimally.

1.2 Composition of PMMA Bone Cement

The PMMA bone cement can uniformly absorb the pressures exerted on the bones due to their optimal elasticity. Because of the strong relationship between bone and cement, the stresses and strain generation at the interface are uniformly distributed. The polymer powder includes benzoyl peroxide (BPO), a free radical polymerization

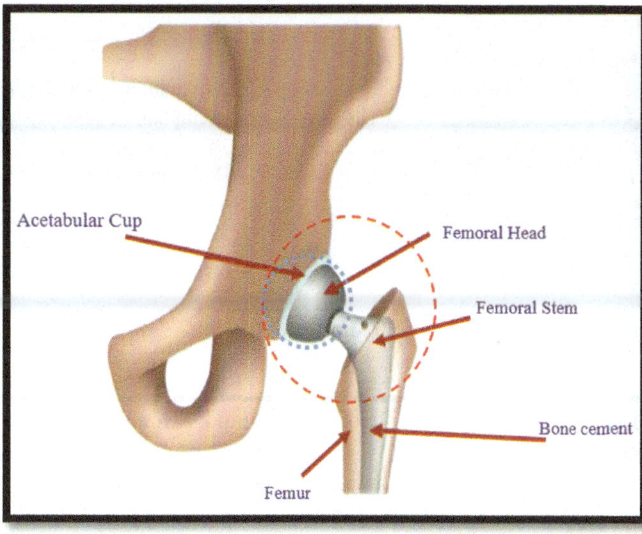

Fig. 1 Prostheses with PMMA bone cement [6]

Polymer	Poly (methyl methacrylate) and/or copolymers of methyl methacrylate
Initiator	Benzoyl peroxide
Radio pacifier	Barium sulphate zirconium dioxide
Antibiotics	Gentamicin sulphate, Clindamycin hydrochloride, Tobramycin, Erythromycin-glucoheptonate and Colistin-methane-sulphonate-sodium
Additives	Dye (chlorophyll) and Plasticiser (di-cyclo-hexyl phthalate)

Fig. 2 Elements of polymer powder of PMMA bone cement

initiator. It can be integrated into the powder or as a component of polymer micro-beads. The primary component in the liquid phase is methyl methacrylate (MMA). However, another methacrylate such as butyl methacrylate can also be found. MMA must be polymerizable; it must have a breakable carbon double bond for it to be utilized in bone cement. An aromatic amine, such as N-dimethyl-p-toluidine, is present in the liquid, which acts as a radical activator. It also includes a storage inhibitor (hydroquinone) that prevents premature polymerization and a colorant like chlorophyll. Figures 2 and 3 show the polymer powder and liquid monomer elements of PMMA bone cement.

Monomer	Methyl methacrylate Butyl methacrylate
Activator	DmpT (*N,N*-dimethyl-*p*-toluidine) DmapE (2-[4-(dimethylamine)phenyl] ethanol)
Inhibitor	Hydroquinone
Additives	Dye (chlorophyll)

Fig. 3 Elements of the liquid monomer of PMMA bone cement

Physical Properties	:	Clear, Colourless liquid, Strong odour
Boilling Point	:	100°C
Density	:	0.943 g/cm^3 at 20°C
Vapour Pressure	:	38 hPa at 20°C
Molecular Weight	:	100 g/mol
Odour threshold	:	0.2 ppm

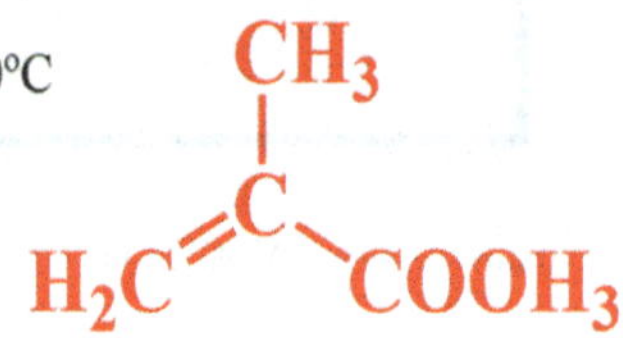

Fig. 4 Properties and chemical structure of MMA [8]

1.2.1 Liquid Monomer to Polymer Granules Ratio

The liquid monomer/polymer powder ratio in most commercially offered bone cement formulas is 1:2 (volume, mL/weight, g). When the 1:2 ratio is reduced, the viscosity of the bone cement paste increases (i.e., lowering the polymer powder content of PMMA bone cement and increasing the MMA liquid monomer content). Commercial cement has a higher powder content to decrease shrinkage caused by a change in density during the conversion of MMA to PMMA during the polymerization process. The powder-to-liquid ratio is estimated to fluctuate by about 0.3 [7].

1.2.2 Reaction of Polymerization

Two primary components of PMMA bone cement include a liquid monomer (Fig. 4) and a polymer powder (Fig. 5). A free radical polymerization (Fig. 6) process produces the finished cement, which takes 8–13 min, dependent on the bone cement blending requirements, environment, and formulation. The cured polymer resulting from this process is PMMA bone cement, which comprises long-chain molecules consisting of thousands of monomers.

Physical Properties	:	Fine powder, Polymer beads
Bead Dia.	:	1-120 μm
Density	:	1.18 g/cm^3
Solubility	:	in Liquid monomer
Molecular Weight	:	0.8 g/mol

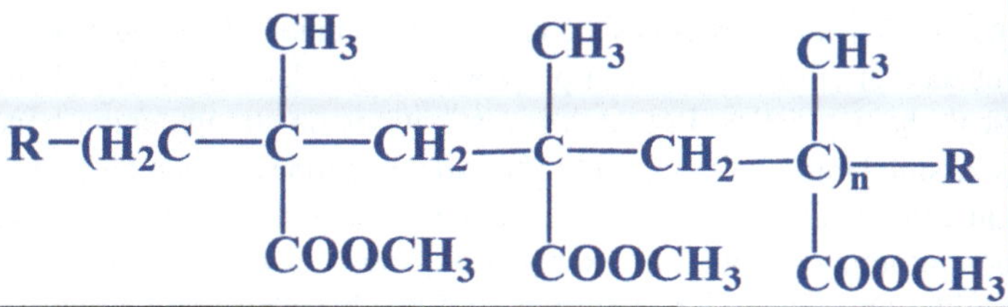

Fig. 5 Properties and chemical structure of PMMA [8]

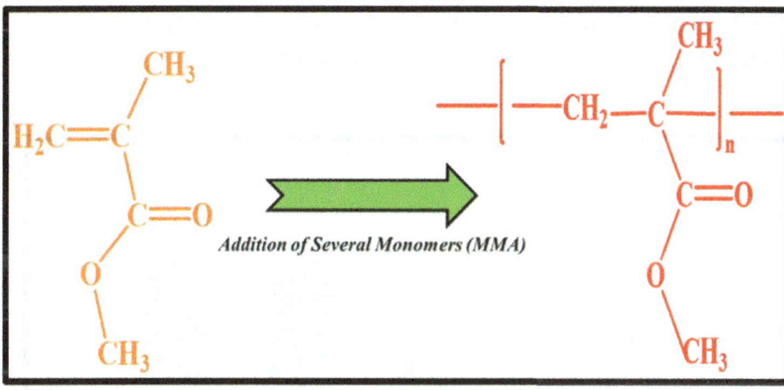

Fig. 6 Polymerization reaction of PMMA [8]

1.3 Application of PMMA Bone Cement

PMMA is also utilized as a plaster material in joint substitution medical procedures to support the bone and prosthetic embed [9]. The numerous utilizations of PMMA in the biomedical are connected particularly to the following territories:

(a) Bone cement for dental and muscular use.
(b) Contact and intraocular focal point.
(c) Screw obsession in bone and vertebrae adjustment in osteoporosis infection.
(d) Filler for various bone holes is increasingly used as an anti-microbial conveyance framework for treating osteomyelitis [10, 11].

2 Additives Used in PMMA Bone Cement

Even though PMMA is generally utilized in orthopedic applications, a few inconveniences, such as removing an auxiliary crack of vertebral bodies, account for lack of mechanical properties [12, 13]. The primary disadvantage is the arthroplasty aseptic loosening which results in short clinical life. This is directly connected to the cement mechanical characteristics, particularly the cement resistance to fracture in the mantle at the cement-prosthesis contact or the cement–bone interface. In recent years, substantial efforts have been proposed by scholars to enhance bone cement mechanical characteristics. Research investigations on strengthening bone cement may be classified into two groups: (1) experimental bone cement design using novel radiopaque agents and (2) commercial bone cement reinforcement using various fillers as reinforcing agents. The exceptional properties of nanoscale materials and the consolidation of nanoscale materials as fillers have been widely investigated as expected solutions for the improvement of the mechanical characteristics of PMMA [14]. Figure 7 shows the mechanical property of cortical bone and other common biomaterials.

The inclusion of water-soluble non-ionic iodine-based radiopacifies such as iohexol (IHX), and iodixanol (IDX) was demonstrated to provide considerably

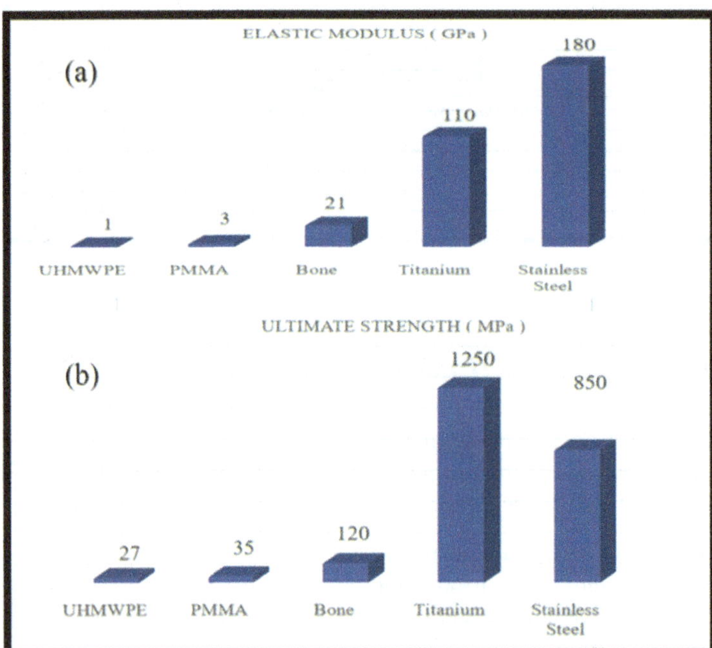

Fig. 7 Mechanical property of cortical bone and other common biomaterials **a** elastic modulus and **b** ultimate strength [15]

greater ultimate tensile strength and strain to failure when it is compared to Palacos® commercial cement at 8 wt.% loading [16]. In contrast to general bio cement containing Barium sulphate ($BaSO_4$) as a radio pacifier, the integration of the triphenyl bismuth (TPB) as a radio pacifier resulted in improved mechanical properties. The findings reveal improved mechanical characteristics at 10 wt.%, such as ultimate tensile modulus and strain rate [17]. The 2 wt.% of multi-wall carbon nanotube (MWCNT) addition increased flexural strength and yield stress [18]. In comparison to the control cement, 10 wt.% Hydroxyapatite (HA) addition increased compression strength and flexural strength [19, 20]. The inclusion of 0.1 wt.% Graphene (G) with silanes enhanced compression strength and flexural strength as equated with control cement [21]. The investigations show that without nanoparticles, 1 wt.% nanoCaCO3 produced improvements in elastic modulus, bending strength, and bending modulus of 10%, 7%, and 8%, respectively [22]. The use of 5 wt.% mesoporous silica nanoparticles (MSNs) increased the flexural strength and compressive strength [23]. The incorporation of Core–shell nanoparticles with a rubbery core of polybutylene acrylate (PBA) and a PMMA shell at a concentration of 10% increased fracture toughness [24]. Incorporation of 0.04 wt.% Graphene oxide (GO) increased flexural and compressive strength in comparison to control cement [25]. They were incorporating Graphene oxide and Chitosan at a 25 wt.% concentration increased the compression and flexural behavior than the control cement [26].

As can be seen from the preceding discussion, nanotechnology can make a significant difference in the growth of bone cement for improved mechanical and physical properties. Table 1 gives a brief detail of different additives enhancing the performances of PMMA bone cement.

Table 1 Improved properties due to different additives

Reinforcement	Quantity (wt.%)	Property improved
Iohexol (IHX) and Iodixanol (IDX)	8	Ultimate tensile strength and strain to failure
Triphenyl bismuth (TPB)	10	Ultimate tensile strain and modulus
Multi-wall carbon nanotube (MWCNT)	2	Flexural strength and yield stress
Hydroxyapatite (HA)	10	Compression strength and flexural strength
Graphene (G)	0.1	Compression strength and flexural strength
Calcium carbonate (nano$CaCO_3$)	1	Elastic modulus, bending strength, and bending modulus
Mesoporous silica nanoparticles (MSNs)	5	Flexural and compressive strength and modulus
Polybutylene acrylate (PBA)	10	Fracture toughness
Graphene oxide (GO)	0.04	Flexural and compressive strength
Graphene oxide and Chitosan	25	Flexural and compressive strength

3 Hydroxyapatite-PMMA Bone Cement Nanocomposite

Collagen, bone mineral, and bone matrix are the three main components of bone. The mineral phase of bone accounts for 60–70% of the total weight and calcium phosphate, similar to hydroxyapatite [27]. In the human body, the nature of hydroxyapatite is biocompatible and bioactive. Since Hydroxyapatite (HA) is osteoconductive, it firmly integrates with the bone and promotes bone formation by bonding directly to the bone, which is advantageous for quick healing. HA has the chemical formulation as ($Ca_{10}(PO_4)_6(OH)_2$) and bone cement increases the combined shear strength at the prosthetic implant interface [28]. Singh et al. [29] demonstrated a comparative investigation for PMMA, the expansion of HA in a PMMA network enhances osteoblast reaction. HA is primarily used in bio reinforcing material for the fabrication of artificial bone because it consists of unique density, stiffness, and bioactivity [30]. HA is remarked as an osteoconductive and biocompatible material that reveals an extracellular reaction to bone formation [31]. It can be used to support the surgical site in a dough state and used as a bone substitute. Many such trials using HA followed, but a bioactive composite cement with high mechanical properties has not been obtained. Because when the wt.% ratio of HA increased above 20, the properties in the mechanical category of the formed cement decreased [32]. Studies illustrated that the combination of hydroxyapatite-reinforced PMMA (PMMA/HA) could become a capable bone cement [33]. The supplement of up to 15 wt.% HA in the powder portion of the PMMA expands the flexural modulus from 2 to 2.5 GPa. Moreover, compared to unreinforced PMMA cement, the option of HA filler showed an improvement in the flexural modulus of approximately 25% [34]. The HA modified biomaterials could be used for bone tissue functions, which requires a 3D scaffold with adequate bone behavioral characteristics, such as mechanical strength, pore size, and osteoconductive, like a bone graft replacement [35]. Because just three fundamental properties are required in artificial bone: (a) mechanical strength, (b) osteoconductive, and (c) biocompatibility. The above findings show that features described above are incorporated in the HA/PMMA based biomaterial to substitute the bone constituents.

The findings of an extensive literature survey remarked that the investigations on the development of HA/PMMA are passing through the initial stage. It requires more attention from the researcher and biomaterial manufacturer side. Still, studies regarding the addition of Hydroxyapatite (HA) into PMMA bone cement have not been thoroughly explored. Present work targets the development of PMMA bone cement modified by HA powder. Three types of composites were prepared using three different weight ratios of 4 wt.%, 10 wt.%, and 16 wt.% of HA. Their mechanical properties like compression strength and flexural strength are analyzed to evaluate the application potential. The FTIR and XRD analysis later investigated the crystal structure behavior and the presence of functional groups. Also, the SEM analysis for fractured surfaces was obtained after mechanical testing to explore the mechanical properties.

4 Materials and Processes

In the fabrication of nanocomposites, Simplex© PMMA bone cement was supplied by M/s. Global Corporation, Ahmedabad, India, and Hydroxyapatite provided by M/s. Nano Research Lab, Jharkhand, India. The concentrated NH_4OH supplied by M/s. Scientific emporium, Gorakhpur, India, and the basic features of the hydroxyapatite are displayed in Table 2.

Composites with various measures of HA were prepared. After mixing HA powder in a concentrated NH_4OH solution thereafter, PMMA in powder form with an equivalent portion by weight of HA was supplemented. The amount of PMMA powder subbed by the hydroxyapatite nanoparticle fillers was 4, 10, and 16%. The proportion of the fluid monomer (M) to the PMMA powder (P) was M/P = 0.5, but the PMMA and HA particles were incorporated into the volume of the powder. By combining the MMA monomer with the PMMA-HA powder, sample preparation was performed. The mixture could be set at room temperature for a one-hour duration. The samples were grouped into four sorts, as mentioned in Table 3, and steps for sample preparation are shown in Fig. 8. The images of the prepared samples are described in Fig. 9a–c.

Table 2 Properties of reinforced hydroxyapatite

Factors	Features
Molecular formula	$Ca_{10}(PO_4)_3(OH)$
Purity	99.5%
Average particle size	20–80 nm
Morphology	Spherical
Molecular weight	1004.6 g/mol

Table 3 Classification of specimens

Sample	PMMA	HA (gm)	Wt.%
Pristine-PMMA	40.0	0	0
HA-PMMA-4	38.4	1.6	4
HA-PMMA-10	36	4	10
HA-PMMA-16	33.6	6.4	16

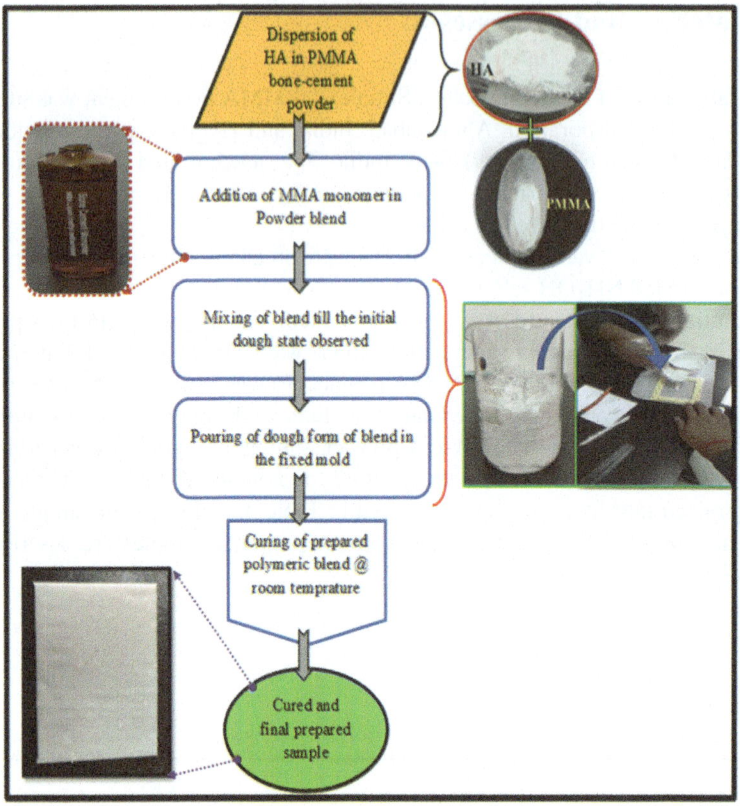

Fig. 8 Flow chart of sample preparation

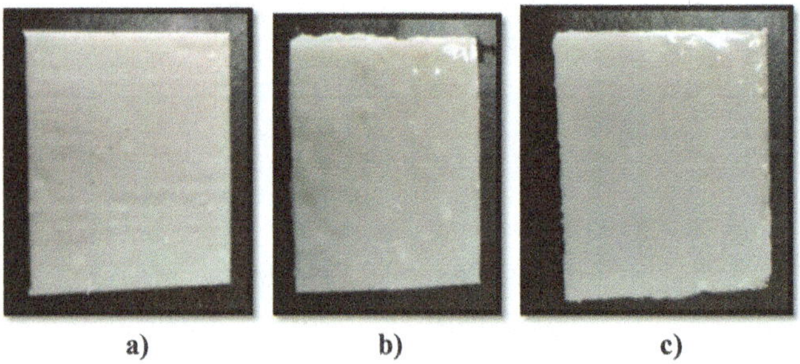

Fig. 9 Prepared sample. **a** HA-PMMA-4. **b** HA-PMMA-10. **c** HA-PMMA-16

5 Crystal Structure Analysis

5.1 *Fourier Transform Infrared Spectroscopy*

FTIR validates the influence of functional groups in the proposed nanocomposite [36]. The nanostructure and elemental analyses of the prepared samples of HA/PMMA were investigated through FTIR. The FTIR BOMEM DA8 spectrometer is a testing-oriented setup with customizable detectors, light sources, and beam splitter. The graph obtained by FTIR analysis indicates the functional groups contained in the PMMA bone cement, HA, and HA-PMMA-10 samples, as shown in Fig. 10. The transmission peaks at 1065 cm^{-1} and 837 cm^{-1} were obtained, which correspond to PMMA bone cement, as found in the previous investigation [37]. At 1048 cm^{-1}, a strong PO_4^{3-} peak indicating the phosphate group emerged, as it had in earlier research [38–40]. As shown in Fig. 10, phosphate-related infrared bands were detected at 1097 cm^{-1} and 962 cm^{-1}. Phosphate member bands may also be seen at 810 cm^{-1} and 757 cm^{-1}. The necessary functional group of nanoparticles in the produced nanocomposite is validated using the observed transmission peaks and comparing them to prior research. This concludes that PMMA bone cement was grafted to HA, as shown by the FTIR spectra.

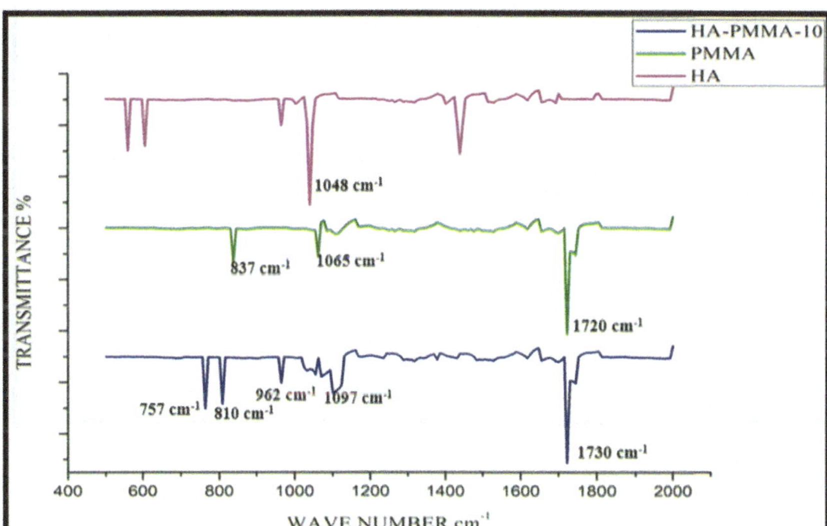

Fig. 10 Infrared spectroscopy

5.2 X-Ray Diffraction

To make use of nanoparticles' properties, it must be evenly distributed into the base material. The chemical composition and homogeneous dispersion of nanomaterials in the resulting composite are validated using X-ray diffraction (XRD). Using a Raguku Model D/max III diffractometer, the crystal structure and phase composition of the generated HA/PMMA nanocomposite were studied. Nickel-filtering was used on the X-ray beam. Cu-Ka1 (lZ0.1504 nm) was irradiated at 40 kV, and 30 mA. Figure 11 shows the acquired peaks of HA, PMMA bone cement, and HA-PMMA-10 nanocomposite in a combined graph. After examining the graph and comparing it to other research, it was discovered that PMMA bone cement exhibits the required peak values [41]. Hydroxyapatite was also found on a polymeric matrix during the test. The peaks of the HA predictor ($2\theta = 25.4°$, 31.4°, and 49°) were identified, and earlier research verified them [26, 42]. The HA nanoparticles are completely distributed in the produced HA-PMMA nanocomposite as per the XRD graph, which shows a similar trend in earlier investigations [43, 44].

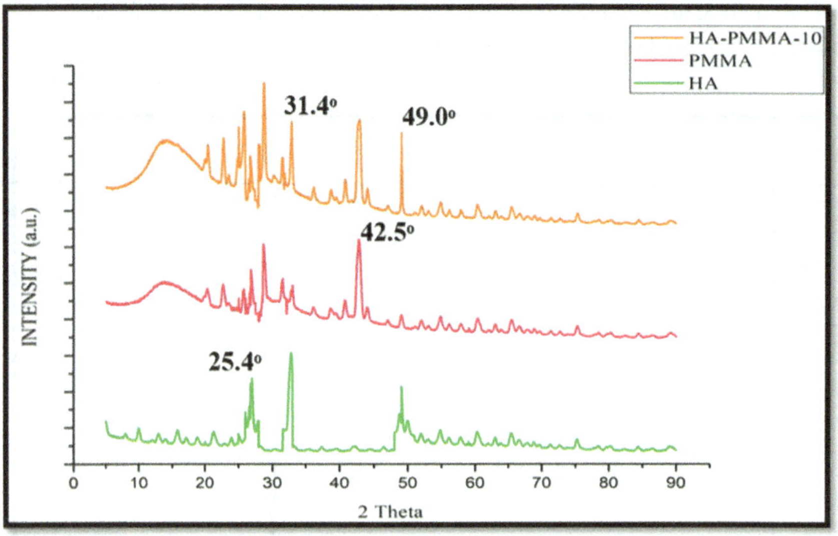

Fig. 11 X-ray diffraction peaks

6 Mechanical Strength Analysis

6.1 Compression Strength Analysis

In applications such as tooth implantation and prosthesis replacement, bone cement compressive strength is essential [45]. To determine the optimal nanomaterial reinforcing amount, measurement of the compressive strength of the produced nanocomposites was obtained. The compressive strength of HA/PMMA nanocomposite and unfilled bone cement was tested. The compression test was carried out according to ISO 5833-2002 [46]. The length and diameter of the specimens used in this study were 12 ± 0.1 mm and 6 ± 0.1 mm, respectively. During the compression study, the load and displacement were observed, and the loading was stopped when an error occurred. For each sample (Table 3), three replications of the test were noted. The add up of nanomaterials significantly expands the compression strength, as remarked in Fig. 12. The strength of the HA-PMMA-10 sample was the greatest.

In contrast to commercially available PMMA bone cement, the compressive strength of the HA-PMMA-10 sample was greater by 21.2% (from 85.3 ± 0.4 to 103.4 ± 0.4 MPa). The compressive strength rises considerably when the wt.% of HA increased up to 10 wt.%, but declined when loaded over 10 wt.%, as indicated in Fig. 12. As previously found in earlier research, agglomeration of HA nanoparticles can cause a decrease in compression strength under heavy loading [20].

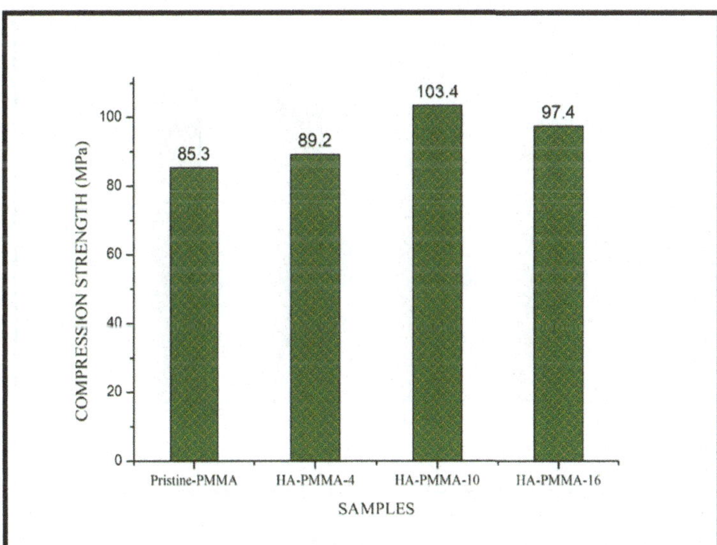

Fig. 12 Compression strength

6.2 Flexural Strength Analysis

Flexural strength becomes a crucial characteristic for applications such as complete hip joint replacement and spine bolt fixation. As a result, flexural property investigation of the produced PMMA-HA nanocomposite is required. The flexural test was carried out according to ISO 5833-2002 [46]. Equation 1 calculates a specimen's flexural strength in a three-point bend test.

$$s = \frac{3 \times P_{max} \times L}{2 \times b \times h^2} \tag{1}$$

where

P_{max} = maximum load faced,
b = width,
h = thickness,
and L = between inner loading points gap (20 mm).

Adding different amounts of nanomaterial can considerably improve the flexural strength (Fig. 13). In comparison to the pristine sample, the flexural strength of the HA-PMMA-10 sample rose by 12.8% (from 63.3 to 71.49 MPa). A decrease in

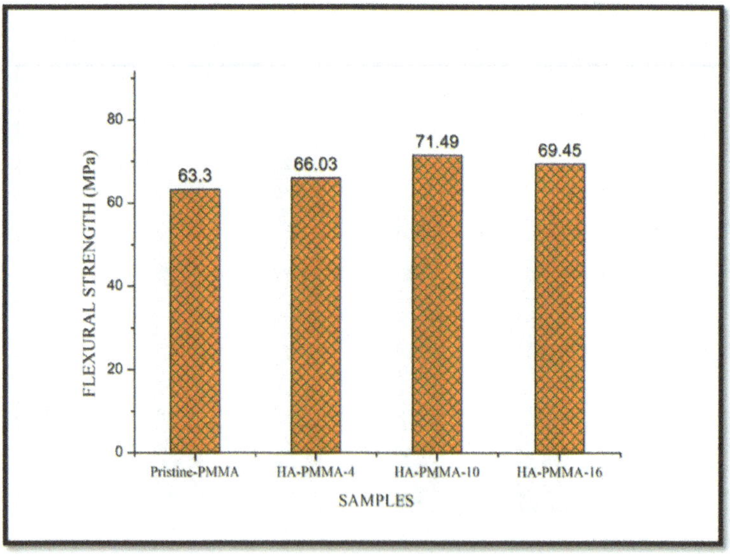

Fig. 13 Flexural strength

flexural strength is also observed when a large amount of HA is used. When hydroxyapatite particles are present in small amounts and are consistently distributed, they function as load carriers, assisting in improving mechanical characteristics. The non-uniform distributions of hydroxyapatite particles can occur at higher loading, resulting in particle aggregation, poor matrix adherence, and decreased mechanical properties [47, 48]. According to the current findings, HA should only be used up to 10% by weight for higher flexural strength.

7 Morphology Analysis

7.1 *Scanning Electron Microscopy Analysis*

The fracture property of bone and related composites such as PMMA bone cement can be studied using the SEM test [49]. The fractured surface was examined to determine the cause of mechanical failure in the nanocomposite. The morphology of the proposed nanocomposite fracture surface was investigated using an XL-30 SEM after the mechanical test. The energy of the electron beam was 10 kV. The fractured specimens were sectioned and sputtered with gold. The produced nanocomposite mechanical property study reveals that the property improves with nanoparticles and that the property decreases as the wt.% increases above 10 wt.% of nanomaterials increases. Figures 14 and 15 indicate the SEM images of all four fractured samples surfaces after the compression and flexural test, respectively at a resolution of 200 μm. The HA-PMMA-4 sample has a rough surface, suggesting poor nanoparticle distribution, whereas the HA-PMMA-10 sample has a smooth surface, showing fine nanoparticle dispersion. As previously reported, SEM inspection of the composite bone cement fracture surface confirmed this claim [48]. The HA dispersion in the HA-PMMA cement cross-section became less uniform as the HA concentration increased, as demonstrated in the HA-PMMA-16 sample. The homogeneous distribution is apparent in the HA-PMMA-10 sample, resulting in excellent mechanical strength [50].

8 Conclusions

This chapter discusses the development and mechanical characterization of HA/PMMA bone cement for orthopedic functions. A cost-effective and durable procedure developed the nanocomposites. The compression and flexural test findings define the feasibility for a cost-effective and improved strength biomaterial product. This study investigates the effect of HA reinforcement weight percent and its supplement in the PMMA for bioproduct fabrication. The findings of the proposed work can be summarized as:

Sample	SEM Images
Pristine-PMMA	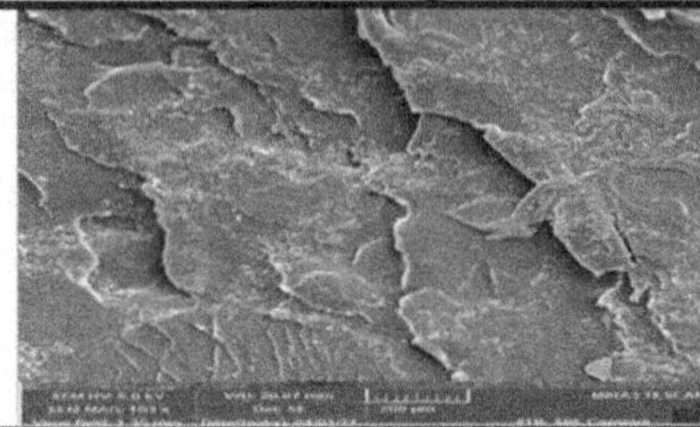
HA-PMMA-4	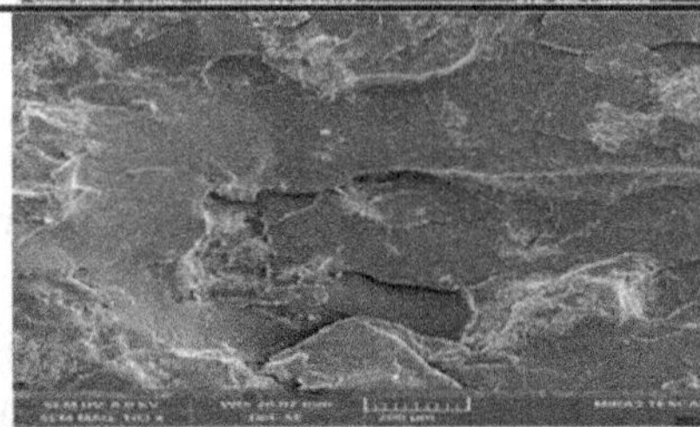
HA-PMMA-10	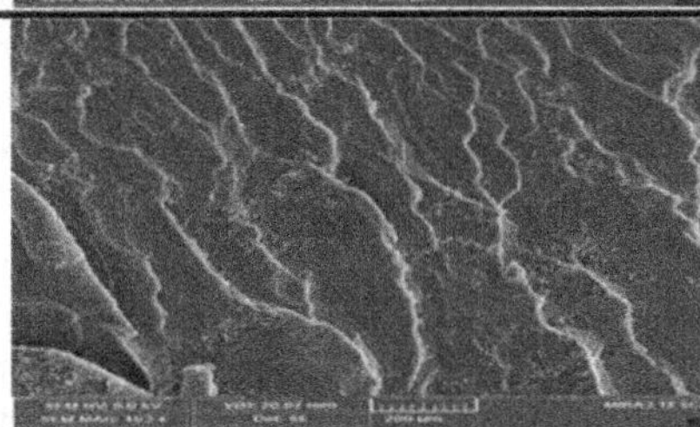
HA-PMMA-16	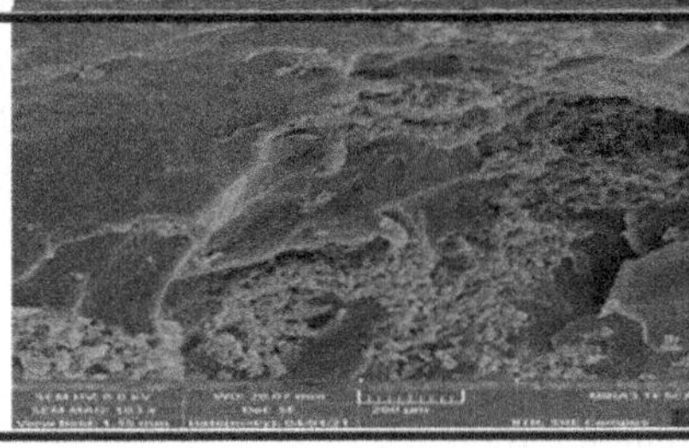

Fig. 14 SEM images of the fractured surface after compression test

- The HA-PMMA bone cement nanocomposite was successfully prepared using a cost-effective procedure. The mechanical characteristics of PMMA bone cement were improved by adding merely 10 wt.% HA.
- In comparison to the pristine PMMA bone cement, the integration of 10 wt.% HA resulted in a 21.2% increase in compression strength.
- At 10 wt.% of HA reinforcement in PMMA based bone cement, the flexural strength of PMMA based bone cement also improved by 12.8%.

Sample	**SEM Images**
Pristine-PMMA	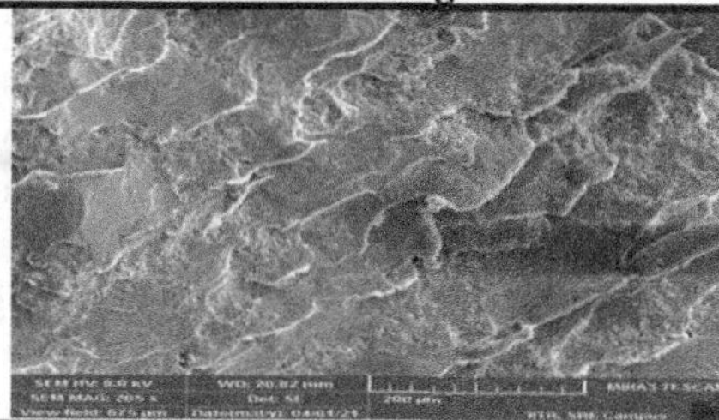
HA-PMMA-4	
HA-PMMA-10	
HA-PMMA-16	

Fig. 15 SEM images of the fractured surface after the flexural test

- By comparing the acquired pattern peaks in XRD and FTIR to prior results, the XRD and FTIR results revealed appropriate dispersion of reinforcement of hydroxyapatite in PMMA bone cement.
- The distribution of HA in the PMMA bone cement matrix at different wt.% may be seen in an SEM picture of the fractured section. The HA-PMMA-16 sample, which has the most significant reinforcement amongst all four samples, has voids and accumulation owing to the non-uniform distribution of HA, resulting in a reduction in mechanical properties.

The current study demonstrates the feasibility of the proposed biopolymer nanocomposites for orthopedic component development. It can also be used for other sectors such as food preservation, tissue engineering, and gene therapy. The discovery of enhanced characteristics such as decent compressive and flexural strength is critical for the long-term viability of prostheses and implants. The inclusion of additional material (HA) substantially improves the desired characteristics. Finding enhanced qualities such as strong compressive and flexural strength is critical for the long-term life of a prosthesis or implant. Furthermore, additional inspections may be carried out as part of the project's progress, necessitating extra analysis and testing. To establish the nanocomposite's overall increase in mechanical properties, mechanical testing such as fracture strength and tensile strength is necessary. The varying weight percentages might potentially be used in further study to establish the best reinforcing value. In the future, further chemical study and some more advanced characterization are required to utilize the proposed biopolymers in society and trade interest.

References

1. Machado C, Sanchez E, Azer SS, Uribe JM (2007) Comparative study of the transverse strength of three denture base materials. J Dent 35:930–933
2. Meng TR, Latta MA (2005) Physical properties of four acrylic denture base resins. J Contemp Dent Pract 6:93–100
3. Anusavice KJ, Shen C, Rawls HR (eds) (2012) *Phillips' science of dental materials*. Elsevier Health Sciences
4. Judet J, Judet R (1950) The use of an artificial femoral head for arthroplasty of the hip joint. J Bone Joint Surg. British volume 32:166–173
5. Charnley J (1960) Anchorage of the femoral head prosthesis to the shaft of the femur. J Bone Joint Surg. British volume 42:28–30
6. Dunne N, Clements J, Wang JS (2014) Acrylic cements for bone fixation in joint replacement. In Joint replacement technology. Elsevier Inc., pp 212–256
7. Meyer PR Jr, Lautenschlager EP, Moore BK (1973) On the setting properties of acrylic bone cement. JBJS 55:149–156
8. Dunne N (2008) Mechanical properties of bone cements. In Orthopaedic bone cements. Woodhead Publishing, pp 233–264
9. Sayeed Z, Padela MT, El-Othmani MM, Saleh KJ (2017) Acrylic bone cements for joint replacement. In Biomedical composites. Woodhead Publishing, pp 199–214
10. Calhoun JH, Mader JT (1989) Antibiotic beads in the management of surgical infections. Am J Surg 157:443–449
11. Kitoh S, Suzuki K, Kiyohara T, Kurita K (1994) Adhesive monomers to dental ceramics. II. Methacrylamide derivatives containing carboxyl and phenyl groups for effective adhesion of calcium metaphosphate ceramic. J Appl Polym Sci 51:2021–2025
12. Çökeliler D, Erkut S, Zemek J, Biederman H, Mutlu M (2007) Modification of glass fibers to improve reinforcement: a plasma polymerization technique. Dent Mater 23:335–342
13. Jiang HJ, Xu J, Qiu ZY, Ma XL, Zhang ZQ, Tan XX, Cui Y, Cui FZ (2015) Mechanical properties and cytocompatibility improvement of vertebroplasty PMMA bone cements by incorporating mineralized collagen. Materials 8:2616–2634
14. Paz E, Forriol F, del Real JC, Dunne N (2017) Graphene oxide versus graphene for optimisation of PMMA bone cement for orthopaedic applications. Mater Sci Eng, C 77:1003–1011
15. An YH, Draughn RA (1999) Mechanical testing of bone and the bone-implant interface. CRC Press

16. Thompson MS, Northmore-Ball MD, Tanner KE (2001) Tensile mechanical properties of polyacetal after one and six months' immersion in Ringer's solution. J Mater Sci - Mater Med 12:883–887
17. Deb S, Abdulghani S, Behiri JC (2002) Radiopacity in bone cements using an organo-bismuth compound. Biomaterials 23:3387–3393
18. Marrs B, Andrews R, Rantell T, Pienkowski D (2006) Augmentation of acrylic bone cement with multiwall carbon nanotubes. J Biomed Mater Res Part A: Off J Soc Biomater, Jpn Soc Biomater, Aust Soc Biomater Korean Soc Biomater 77:269–276
19. Zebarjad SM, Sajjadi SA, Sdrabadi TE, Sajjadi SA, Yaghmaei A, Naderi B (2011) A study on mechanical properties of PMMA/hydroxyapatite nanocomposite. Engineering 03:795–801
20. Phakatkar AH, Shirdar MR, Qi ML, Taheri MM, Narayanan S, Foroozan T, Shokuhfar T (2020) Novel PMMA bone cement nanocomposites containing magnesium phosphate nanosheets and hydroxyapatite nanofibers. Mater Sci Eng, C 109:110497
21. Paz E, Ballesteros Y, Forriol F, Dunne NJ, Del Real JC (2019) Graphene and graphene oxide functionalisation with silanes for advanced dispersion and reinforcement of PMMA-based bone cements. Mater Sci Eng, C 104:109946
22. Hill J, Orr J, Dunne N (2008) In vitro study investigating the mechanical properties of acrylic bone cement containing calcium carbonate nanoparticles. J Mater Sci - Mater Med 19:3327–3333
23. Slane J, Vivanco J, Meyer J, Ploeg HL, Squire M (2014) Modification of acrylic bone cement with mesoporous silica nanoparticles: effects on mechanical, fatigue and absorption properties. J Mech Behav Biomed Mater 29:451–461
24. Gutiérrez-Mejía A, Herrera-Kao W, Duarte-Aranda S, Loría-Bastarrachea MI, Canché-Escamilla G, Moscoso-Sánchez FJ, Cauich-Rodríguez JV, Cervantes-Uc JM (2013) Synthesis and characterization of core-shell nanoparticles and their influence on the mechanical behavior of acrylic bone cements. Mater Sci Eng, C 33:1737–1743
25. Khan AA, Mirza EH, Mohamed BA, Alharthi NH, Abdo HS, Javed R, Alhur RS, Vallittu PK (2018) Physical, mechanical, chemical and thermal properties of nanoscale graphene oxide-poly methylmethacrylate composites. J Compos Mater 52:2803–2813
26. Tavakoli M, Bakhtiari SSE, Karbasi S (2020) Incorporation of chitosan/graphene oxide nanocomposite in to the PMMA bone cement: physical, mechanical and biological evaluation. Int J Biol Macromol 149:783–793
27. Sun L, Berndt CC, Gross KA, Kucuk A (2001) Material fundamentals and clinical performance of plasma-sprayed hydroxyapatite coatings: a review. J Biomed Mater Res: Off J Soc Biomater, Jpn Soc Biomater, Aust Soc Biomater Korean Soc Biomater 58:570–592
28. Kwon SY, Kim YS, Woo YK, Kim SS, Park JB (1997) Hydroxyapatite impregnated bone cement: in vitro and in vivo studies. Bio-Med Mater Eng 7:129–140
29. Singh MK, Shokuhfar T, de Almeida Gracio JJ, de Sousa ACM, da Fonte Fereira JM, Garmestani H, Ahzi S (2008) Hydroxyapatite modified with carbon-nanotube-reinforced poly(methyl methacrylate): a nanocomposite material for biomedical applications. Adv Func Mater 18:694–700
30. Cheang P, Khor KA (2003) Effect of particulate morphology on the tensile behaviour of polymer–hydroxyapatite composites. Mater Sci Eng, A 345:47–54
31. Marino X (2005) Functional fillers for plastics. WILEY-VCH Verlag GmbH & Co.
32. Sogal A, Hulbert SF (1992) Mechanical properties of a composite bone cement: polymethylmethacrylate and hydroxyapatite. Bioceramics 5:213–224
33. Serbetci K, Korkusuz F, Hasirci N (2004) Thermal and mechanical properties of hydroxyapatite impregnated acrylic bone cements. Polym Testing 23:145–155
34. Vallo CI, Montemartini PE, Fanovich MA, Porto Löpez JM, Cuadrado TR (1999) Polymethylmethacrylate-based bone cement modified with hydroxyapatite. J Biomed Mater Res 48:150–158
35. Raval JP, Joshi P, Chejara DR, Disher IA (2008) Fabrication and applications of hydroxyapatite-based nanocomposites coating for bone tissue engineering. In Applications of nanocomposite materials in orthopedics. Elsevier, pp 71–82

36. Schmitt J, Flemming H-C (1996) FTIR spectroscopy. Microbially influenced corrosion of materials. Springer, Berlin Heidelberg, pp 143–157
37. Pahlevanzadeh F, Bakhsheshi-Rad HR, Ismail AF, Aziz M, Chen XB (2019) Development of PMMA-Mon-CNT bone cement with superior mechanical properties and favorable biological properties for use in bone-defect treatment. Mater Lett 240:9–12
38. Balamurugan A, Kannan S, Selvaraj V, Rajeswari S (2004) Development and spectral characterization of poly(methyl methacrylate)/hydroxyapatite composite for biomedical applications. Trends Biomater Artif Organs 18:41–45
39. Chou PM, Mariatti M (2010) The properties of polymethyl methacrylate (PMMA) bone cement filled with titania and hydroxyapatite fillers. Polym Plast Technol Eng 49:1163–1171
40. Quan C, Tang Y, Liu Z, Rao M, Zhang W, Liang P, Wu N, Zhang C, Shen H, Jiang Q (2016) Effect of modification degree of nanohydroxyapatite on biocompatibility and mechanical property of injectable poly(methyl methacrylate)-based bone cement. J Biomed Mater Res Part B Appl Biomate 104:576–584
41. Tan H, Guo S, Yang S, Xu X, Tang T (2012) Physical characterization and osteogenic activity of the quaternized chitosan-loaded PMMA bone cement. Acta Biomater 8:2166–2174
42. Venkatesan J, Qian ZJ, Ryu B, Ashok Kumar N, Kim SK (2011) Preparation and characterization of carbon nanotube-grafted-chitosan—natural hydroxyapatite composite for bone tissue engineering. Carbohyd Polym 83:569–577
43. Asgharzadeh Shirazi H, Ayatollahi MR, Beigzadeh B (2016) Preparation and characterisation of hydroxyapatite derived from natural bovine bone and PMMA/BHA composite for biomedical applications. Mater Technol 31:448–453
44. Kang IG, Park CI, Lee H, Kim HE, Lee SM (2018) Hydroxyapatite microspheres as an additive to enhance radiopacity, biocompatibility, and osteoconductivity of poly(methyl methacrylate) bone cement. Materials 11:258
45. O'Dowd-Booth CJ, White J, Smitham P, Khan W, Marsh DR (2011) Bone cement: perioperative issues, orthopaedic applications and future developments. J Perioper Pract 21:304–308
46. International Standard ISO 5833 Implants for Surgery-Acrylic Resin Cements Implants Chirurgicaux-Ciments à Base de Résine Acrylique ISO 5833:2002(E).
47. Morejón L, Mendizábal AE, García-Menocal JAD, Ginebra MP, Aparicio C, Mur FJG, Marsal M, Davidenko N, Ballesteros ME, Planell JA (2005) Static mechanical properties of hydroxyapatite (HA) powder-filled acrylic bone cements: effect of type of HA powder. J Biomed Mater Res Part B Appl Biomater 72:345–352
48. Rao M, Su Q, Liu Z, Liang P, Wu N, Quan C, Jiang Q (2014) Preparation and characterization of a poly(methyl methacrylate) based composite bone cement containing poly(acrylate-co-silane) modified hydroxyapatite nanoparticles. J Appl Polym Sci 131(15)
49. Boyde A, Maconnachie E, Reid SA, Delling G, Mundy GR (1986) Scanning electron microscopy in bone pathology: review of methods, potential and applications. Scan Electron Microsc 1986:31
50. Aram E, Ehsani M, Khonakdar HA, Abdollahi S (2019) Improvement of electrical, thermal, and mechanical properties of poly(methyl methacrylate)/poly(ethylene oxide) blend using graphene nanosheets. J Thermoplast Compos Mater 32:1176–1189

Conceptual Design for Innovation: Process and a Knowledge-Based Approach

Xin Guo, Yiwei Jiang, and Ying Liu

The traditional design methods mainly focus on product or technical system themselves, but seldom recognize and plan design process from creative cognition approach point of views. This chapter introduces the crucial initial stage of product innovation design. It focuses on a method to innovation design and knowledge-based approach that leads the user through the design process, helps identify critical issues, and proposes new conceptual schemes for specific configurations. The definition and types of innovative conceptual design are introduced, and the factors that affect individual creativity are analyzed. A process of innovative conceptual design is proposed. Based on the characteristics of product design, incorporating cognitive psychology, knowledge-based approach, and information technology into design technology. Based on the cognizing process, the design activities are defined, knowledge is recommended, and then, suitable creative solutions are evaluated and selected.

X. Guo · Y. Jiang
School of Mechanical Engineering, Sichuan University, Chengdu 610065, China
e-mail: guoxin@scu.edu.cn

Y. Jiang
e-mail: jiang_yw@stu.scu.edu.cn

Y. Liu (✉)
Institute of Mechanical and Manufacturing Engineering, School of Engineering, Cardiff University, Cardiff CF24 3AA, UK
e-mail: LiuY81@Cardiff.ac.uk

P. Kyratsis et al. (eds.), *Advances in Product Design Engineering*,
Management and Industrial Engineering,
https://doi.org/10.1007/978-3-030-98124-2_8

1 Introduction and Definition

Concept design process is a solution process that analyzes functional requirements to generate product solutions, which are innovative, complex, and cognitive. It determines 70% of the total product life cycle cost [1]. At this stage, the main task is to generate conceptual solutions through demand analysis and functional solution to meet customer expectations and functional requirements [2]. As the key stage of product innovation design, the conceptual design presents the characteristics of uncertainty, multi-solution, and cross-domain, which is the focus of current design methodology research.

The product concept design process has the characteristics of large design freedom, relatively few constraints on designers, and strong uncertainty [3]. Many researchers all over the world have done a great deal of work from different perspectives of requirements analysis, functional decomposition, and knowledge reasoning [2, 4]. With the increasing requirements for product customization and the wide application of new technologies such as the Internet and big data, the requirements for innovation in the conceptual design process have also undergone new changes.

What is innovation? There are many definitions of it. The common definition from Cambridge Learner's Dictionary is—

> *Innovation is a new idea or method that is being tried for the first time, or the use of such ideas or methods.*

According to the psychologist Sternberg [5], innovation is related to a person's knowledge, thinking styles, personality, motivation, and environmental context. Li et al.[6] proposed a design creativity (DC) model for product innovation, that expressed as:

$$DC = f(K, TS, DM, ST, O)$$

where K is knowledge, TS is thinking styles, DM is design methods, ST is supporting tools and O is other factors. These other factors could form an important component of innovative design, which leads to uncertainty and ambiguity in the design process. On this basis, the innovative concept design can be defined as follows:

Innovative conceptual design is a dynamic process that stimulates the creativity of designers, and uses new technologies, principles, and methods to design novel conceptual schemes. In this process, user's functional requirements are satisfied through knowledge reasoning and deployment. It starts with requirement acquisition and ends with the generation of a new conceptual scheme.

Innovative conceptual design can be divided into the following four types according to different design goals [6].

1. Issue-oriented innovative design

The purpose of the design is to improve the performance of an existing product or system and solve the issues that affect the main function of the product or system. The strategy taken is to solve the minimized issues, which means resolving the system conflicts by only making some incremental changes (first-order change) of the system each time.

2. Function-oriented innovative design

The purpose of the design is to design a new product or system. In order to avoid the improvement of main functions of the existing product or system reaching its limitation, the designers should jump out of the existing system to find better ways of possible existence to reach it. The strategy taken here is to solve the maximized issue and change working principles (second-order change) of the system.

3. Product-oriented innovative design

The purpose of the design is to explore and improve the product's sub-function in the situation that the main function does not need to be changed at the moment. The strategy taken here is to improve the product itself using the methods of functions-follow-forms and it is mostly used in the condition that the product has a good reputation on the market.

4. Form-oriented innovative design

The purpose of the design is to meet customers' cultural and mental needs through the way of changing the shape, format, or appearances of a product when the main function and the corresponding implementing principle of the product are ascertained. The strategy taken here is the utilization of the nonlogical thinking method, which creates novel design concepts by an association of ideas, and/or random inspiring methods.

Innovative conceptual design has become the development trend in the field of product design research because it can effectively meet customers' requirements for product diversification and improve the competitiveness of the product market.

To improve the efficiency of conceptual solutions and reduce the uncertainty caused by empirical decisions and knowledge limitations, in this chapter, the definition and types of innovative conceptual design are introduced, and the factors that affect individual creativity are analyzed. A process of innovative conceptual design with good operability is introduced.

2 Product Conceptual Design Process

2.1 Main Features

The product concept design process is a solution process that analyzes functional requirements to generate product solutions, which are innovative, complex, and cognitive [7]. An executable conceptual design model usually has the following four features.

- It can accurately analyze and classify design tasks to clarify design issues and facilitate designers to determine design directions.
- The product solution process can be guided by the solution strategy and design methods, and the designers are assisted and motivated by the knowledge resources according to different design contexts.
- The whole conceptual design process should ensure that the designers' thinking is fully dispersed, while ensuring timely convergence of design thinking at the necessary moment, so as to guide the flow of creative thinking of designers.
- The conceptual design process can be evaluated by feedback in a timely manner, and the design direction can be adjusted in an iterative manner to obtain the optimal design solution.

In the early stages of design, designers have only an incomplete and relatively vague mental representation of the upcoming design, so a conceptual design model needs to be established in order to accomplish the following tasks.

- Helps to understand the cognitive process of designers in the conceptual design process.
- Helps to understand the information of the conceptual design process and how it is delivered.
- Helps to clarify the constraints of each step of the cognitive operation process.

2.2 The Multi-Stage Mapping Processes of Product Design

The conceptual design process is a multi-stage mapping process including customer requirements domain, functional structures' domain, conceptual scheme domain, and scheme evaluation domain shown in Fig. 1. In this process, the customer requirements domain is the target descriptive set of design products which meets customer requirements. The functional structures' domain is the physical structures' set produced from a certain scientific principle. The conceptual scheme domain is the satisfied conceptual schemes' set in which the structures meet the design constraints after various extending conversions [8]. The scheme evaluation domain is the evaluations' set of the conceptual program from users and designers that determines the optimization direction and evolution result of the conceptual scheme.

1. The customer requirements domain—Analysis

The design task analysis and the design issue analysis are included in the customer requirements domain. The design task analysis mainly refers to the requirement analysis and sometimes to the product analysis. In order to generate the product solutions, the functional requirements are analyzed by the product conceptual design processes [7]. There are three goals for requirement analysis: (1) User's descriptions are interpreted using questionnaire, conversation method, and the 5W1H method.

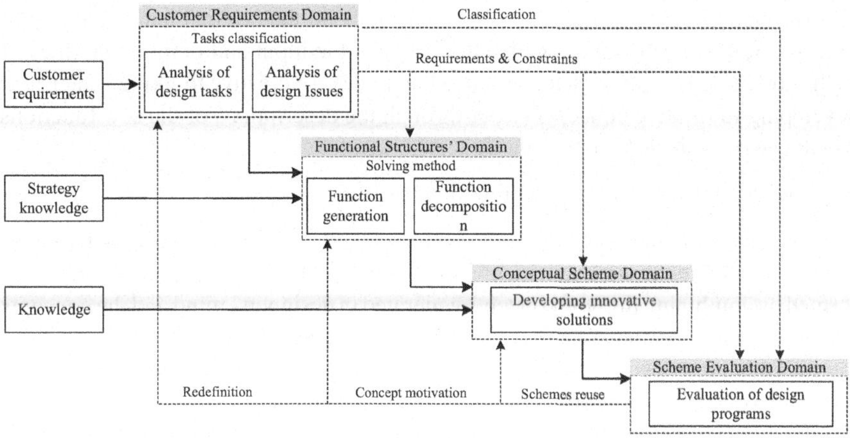

Fig. 1 The multi-stage mapping processes of product design

(2) The descriptions are transformed into concise and functional requirements. (3) According to the requirements, the macro-design tasks are transformed into the operational design issues.

The awareness and understanding of the initial conditions (the goals of the tasks or the issues of the tasks components) are included in the design issue analysis. The exploration of the requirements and constraints that need to be met in the design are also included in it. During the design process, constraints and requirements may be changed due to the changes in issues representations. The evaluation criteria are determined according to the design goals. It is a part of the issues analysis.

2. The functional structures' domain—Decomposition

Decomposition mainly refers to a behavior of product function (generation or decomposition). In order to achieve the goal of product design, a clear definition of the product functionality in conceptual design is needed. For products, the function can be understood as efficacy, which is related to, but not the same as, the concepts of use, capability, and performance. The computable physical parameters are transformed from the subjective functional concepts by generally using the Function-Behavior-Structure (FBS) method. The FBS method is also used to analyze and evaluate the physical structure of the concepts.

There are three features in the function: (1) Function is useful for designers to open up their minds and conceive the innovative design. It is an abstract behavior; (2) Function is the embodiment of the interaction process between the products and the environment, and it can be reflected only in the dynamic process of the products; (3) Functions are variable because it can be influenced by the variability of external inputs of energy, material, and information.

Total function can be decomposed into simple sub-functions. It is a proven method to find the working principles of the products. The product contains the ability of functional requirements can also be extended by this behavior used some methods

[1]. Through this decomposition, it is easier to obtain the original understanding of each sub-function since the relationship between the input and output quantities of each sub-function is more clear. It is also the process of creative thinking. The last is developing innovative concepts easier through the descriptions of the sub-functions which can open up designers' minds.

3. The conceptual scheme domain—Composition

Composition is the behavior that develops innovative design concepts based on sub-functions into design schemes. That is, the preliminary design schemes are formed by combining the new concepts generated in designers' mind and the concepts generated in the previous design cycle. The design schemes will be further translated into more mature design solutions. It can be assumed that the technical or physical conflict has occurred when designers arrive at a design scheme that does not match the function requirements [9]. The conflict can be solved by using the Theory of Inventive Issue Solving (TRIZ). The composition is not only a simple superposition and combination of concepts but also a generation of design scheme in the continuous iteration of issues and schemes space.

4. The scheme evaluation domain—Evaluation

There are various methods for design evaluation decisions and the key for them is to determine the evaluation index system, assign weights and calculate the total value of the design. Through two design methods, the conceptual schemes are filtered and iterated by designers. The AHP method is often used to rank the importance of solutions and Axiomatic Design Theory (AD) is used as a criterion to evaluate good designs for designers to filter or iterate on solutions. The evaluation is not only a scientific analysis and summary of the design scheme, but also an improvement and refinement of the design scheme. Determine the evaluation index system, assigning weights and calculate total value are the keys of evaluation. As an exploratory cognitive process, designers can ensure the relevance, usefulness, and suitability of the design schemes through evaluation. Relevance and usefulness are determined by design requirements and constraints, while the suitability is determined by design criteria.

3 The Process Model of Innovative Conceptual Design

Under normal circumstances, the process model of conceptual design includes two key technologies, namely, the expression of design knowledge and the solution of functions, which have received constraints from customer requirements. In order to clearly describe the process of innovative conceptual design, the model in this chapter will discuss the design process from three aspects: issue space, knowledge space, and solution space. The cognitive activities of designers between these three spaces constitute the information flow of the entire product issue solving process. While traditional concept design mainly completes the issue-solution process, it is important to further introduce knowledge into the process to assist designers in generating innovative solutions more efficiently. Therefore, an innovation-oriented product concept design process model is needed, which establishes the iterative mapping relationship between issue space, knowledge space, and solution space from the perspective of design element mapping and reasoning, and builds a product innovation workflow that conforms to the design law and designer's cognitive law with the corresponding mapping process as a guide and the support of relevant issue analysis, solution strategy, knowledge base, and other resources. The mapping process is guided by relevant issue analysis, solution strategies, knowledge base, and other resources to build a product innovation workflow that conforms to design laws and designer cognition laws. A conceptual design model for product innovation is shown in Fig. 2.

In this model, the issue space, design knowledge space, and design solution space are parallel to each other (Fig. 2A). Each of them contains the corresponding design elements and cognitive activities (Fig. 2B). The conduct and iteration of design

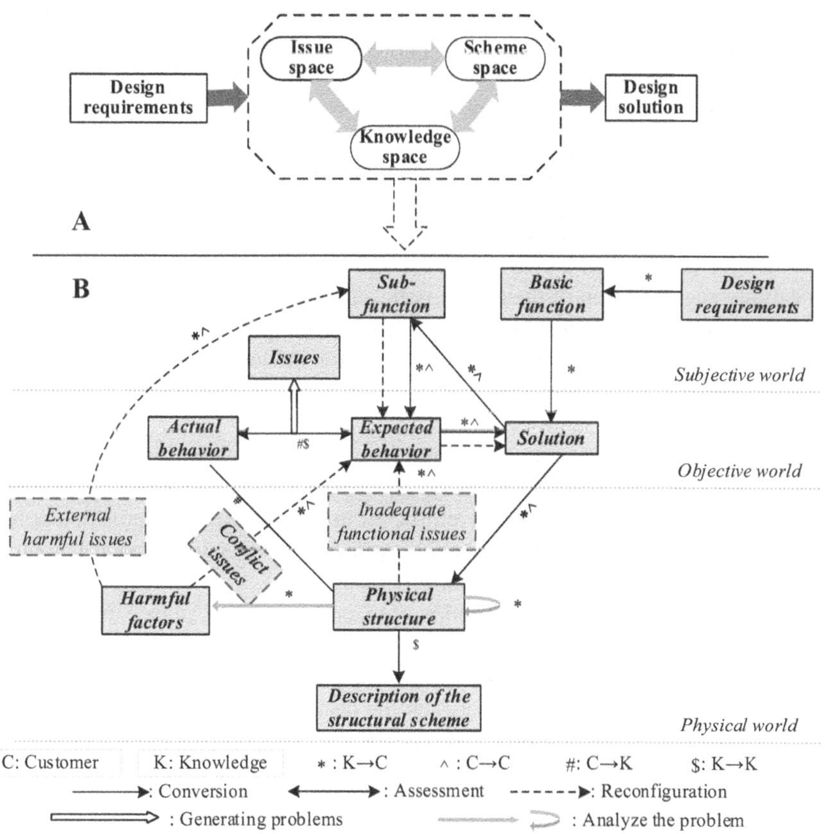

Fig. 2 The process model of innovative conceptual design

activities reflect the changing state of the three spaces. The evolution and mapping processes of the three spaces are explained by four macro processes, which are introduced as follows:

1. Requirements—issue

It can be divided into function-driven innovative design strategies and issue-driven innovative design strategies according to the different starting points of mapping. The function-oriented requirements and the issue-oriented requirements are reflected through two mapping strategies, then the design issue information or design solution information is obtained through the analysis of design requirements.

2. Knowledge—issue—scheme

Three steps are consisted in the process: (1) Part of the issues are discovered or identified through the explorations of the issue space that are supplemented by the knowledge and experience owned by the designer and the "standard questionnaires of mental pre-defined (Default)". (2) The relevant information is selected and checked through the key issues and the "interesting points" that are highlighted with an appropriate perspective. It can be implemented by using relevant knowledge to structure the issues. (3) Process 3 is evolved through a temporary scheme that is obtained through the issues solving with the knowledge used by the designer.

3. Knowledge—scheme—issue

Three steps are consisted in the process: (1) The temporary schemes entered by process 2 are evaluated under the influence of empirical knowledge and external knowledge; (2) Process 2 is evolved through the standardized representation and scalability segmentation of the design issues that are helped by the relevant knowledge. It will happen when the logical states of the temporary schemes are not clear; (3) Process 4 is evolved through the schemes when the temporary schemes are clear.

4. Scheme—knowledge

A scheme essentially contains a solution with a clear state of true or false logic. They can all be used as new knowledge to expand the knowledge space. Process 2 and process 3 are a circular iterative process in it. A complete innovation solving activity is constituted through the iterative evolution between issue space, knowledge space, and scheme space of these four processes.

The method to promote the evolution of macro processes can be expressed as the mapping processes of design elements. Depending on different starting point of the mapping processes, it can be classified as function or issue-driven mapping. There are fourteen mapping processes between the design elements, which are shown in Table 1.

Table 1 Mapping processes of design elements in the model

Mapping process	Description
1. Design requirements map to basic functions	*The requirements or goals are transformed into basic functions* (K → C: Assisted Conversion of K_e)
2. Basic functions map to schemes	*The "temporary work structures or schemes" for the overall functions are Imagine* (K → C: Assisted imagination of K_e)
3. Schemes map to sub-functions	*The sub-functions are formed through the imagined "temporary work structures or schemes"* (K → C: Assisted decomposition of K_e C → C: Limit subdivision)
4. Sub-functions map to expected behaviors	*The sub-functions are converted into the expected behaviors, or be expressed abstractly* (K → C: Assisted abstraction of K_e/K_E C → C: Extended subdivision)
5. Expected behaviors map to schemes	*Get the "temporary" objective structures due to the desired behaviors* (K → C: $K_e/K_E/K_M/K_p$ C → C: Extended subdivision)
6. Schemes map to physical structures	*The physical structures are converted from "temporary" objectives* (K → C: transformation of K_C/K_p C → C: Limit subdivision)
7. Physical structures map to actual behaviors	*The actual behaviors of the physical structure are analyzed* (C → K: Analyze logical state)
8. Mutual mapping of actual behaviors and expected behaviors	*The actual behaviors and expected behaviors are Evaluated. It will enter the issue analysis stage if the structural schemes are not accepted* (C → K: Determining the logical state K → K: If accepted)
9. Physical structures map to themselves	*The physical structures are analyzed to make sure whether it causes the lack of functions* (K → C: Analysis of K_e/K_M)
10. Physical structures map to harmful factors	*The physical structures are analyzed to make sure whether it generates external or internal harmful factors* (K → C: Analysis of K_e/K_M)
11. Physical structures map to expected behaviors	*The objective structures are reconstructed from the synthesis of expected behaviors due to the issue of insufficient functions* (K → C: Reconstruction of K_e/K_M C → C: Limit / extended subdivision)

(continued)

Table 1 (continued)

Mapping process	Description
12. Harmful factors map to expected behaviors	*The objective structures are reconstructed from the synthesis of expected behaviors due to the issue of internal conflicts* (K → C: Reconstruction of K_e/K_M C → C: Limit / extended subdivision)
13. Harmful factors map to sub-functions	*New effects or new expected behaviors are generated by the "substances" that are introduced due to the external harmful issue* (K → C: $K_e/K_E/K_M/K_p$ C → C: Extended subdivision)
14. Physical structures map to the description of the structural schemes	*If the structure schemes are accepted, then it can be used for future activities by being described and recorded* (K → K: Expand the knowledge space)

4 Knowledge Reasoning Method for Innovative Conceptual Design

Conceptual design is a highly creative activity that involves design cognition, innovative methods, and the transfer of multidisciplinary knowledge. An effective knowledge organization and reasoning method help to recommend suitable functional knowledge to the designer during the conceptual design process. If some knowledge were connected to the sub-functions of decomposition, it would help to obtain further insights into the conceptual design stage. In this chapter, we define this kind of knowledge as functional knowledge. The conceptual design model for product innovation introduced in the third section focuses on "when to use knowledge", and this section answers the question "how to reason about functional knowledge".

4.1 Knowledge Organization and Mapping Process

Knowledge-based reasoning in conceptual design matches and operates data by using knowledge-based rules and deriving conclusions by applying certain reasoning methods [10]. A concept-knowledge theory proposed by Hatchuel and Weil is a traditional method used to describe the process of design thinking [11]. It has had a significant influence in guiding the study of knowledge flow in the solution process used to design innovation conceptual schemes.

Knowledge reasoning is a kind of knowledge organization and matching technology that associates suitable functional knowledge with design tasks in different stages of innovative conceptual design. Designers are users of knowledge and decision-makers in product design activities. The effect of knowledge reasoning depends on the matching accuracy between knowledge and design activities. To accurately push the knowledge required for innovative conceptual design, mapping modes and matching algorithms are introduced.

Based on the process model of innovative conceptual design, knowledge reasoning mainly occurs in the process of decoupling conceptual schemes, including customer requirements identification, functional feature decomposition, and functional knowledge matching. Pairwise mapping has become the following relationships [12]:

- "Requirement-function" mapping: mapping customer requirements to functional features such as engineering features, part features, process features, and production features, and the designer understands "what needs to be designed".
- "Function-knowledge" mapping: based on the selection and combination of functional knowledge, transform "what needs to be designed" into "how to do", and realize the generation of conceptual schemes.

Customer requirements and conceptual schemes are closely related, and knowledge connects the two through reasoning and mapping. Innovative conceptual design requires the integration and transfer of multi-domain knowledge. Ontology is an effective method of organizing knowledge. It not only provides a representation of a shared conceptualization of a particular domain that supports communication between users and applications, but it also captures knowledge about concepts in the domain and relationships between these concepts [13]. It helps knowledge reasoning and data integration. By categorizing customer requirements, functional feature and functional knowledge, and defining the relationship between them, an ontology tree structure is shown in Fig. 3. On this basis, the semantic annotation and ontology expression of customer requirement, functional feature and functional knowledge can be established through the Protégé system. It is an environment for knowledge-based systems development.

In the process of innovative conceptual design, a requirement ontology database ($DR = DR_1, DR_2, \ldots, DR_n$) is established based on customer requirements. A similar FF_i will be retrieved from the functional feature ontology database ($FF = FF_1, FF_2, \ldots, FF_n$) to output functional features that meet customer requirement DR_i. Similarly, search for knowledge FK_i similar to FF_i in the functional ontology database ($FK = FK_1, FK_2, \ldots, FK_n$), and then push it to designers to stimulate creative thinking and generate conceptual scheme. Innovative conceptual design is a process of iterative and gradual improvement. In this process, customer requirement, functional feature and functional knowledge are constantly superimposed and fed back, so that the innovative concept scheme is gradually improved. The specific process is shown in Fig. 4.

- Step 1: The functional features, such as material, control, transmission, drive, structure, etc., are expressed based on the ontology, and the functional features ontology database is established.

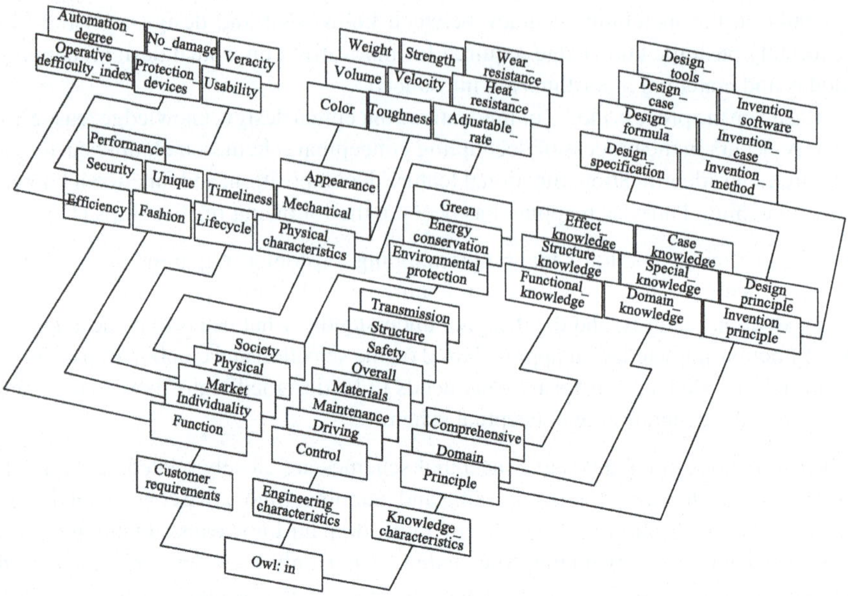

Fig. 3 Ontological relations of innovative conceptual design knowledge

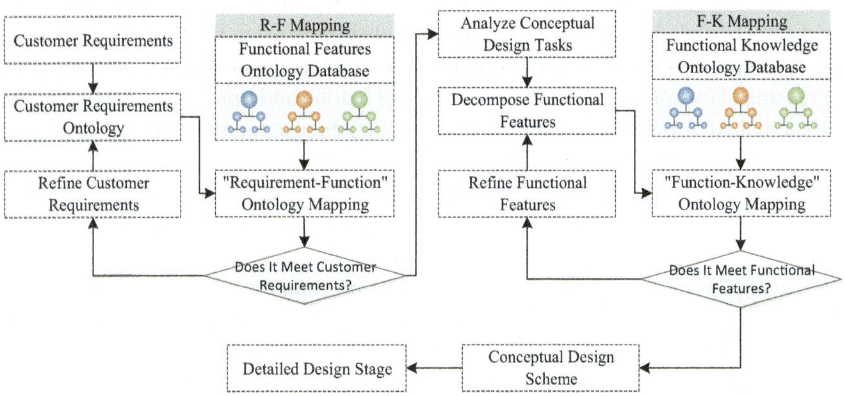

Fig. 4 The mapping process of ontology-based functional knowledge

- Step 2: Functional knowledge such as overall function, usability, structural characteristics, and design principles is expressed based on the ontology, and a functional knowledge ontology database is established.
- Step 3: Customer requirements are input, such as product quality, price, function, performance, appearance, service life, etc., and are expressed normatively based on the ontology.

- Step 4: "Requirement-function" mapping. The designer analyzes and judges whether the mapping result meets the customer's requirements. If it is satisfied, output the corresponding function feature, otherwise, return to step 3 to further refine customer requirements.
- Step 5: According to the function features, the next design task is clarified, and the specific design activities are divided.
- Step 6: "Function-knowledge" mapping. The designer analyzes the mapping results and judges whether the acquired functional knowledge meets the requirements of function features. If it is satisfied, output the conceptual scheme, otherwise, return to step 5 to further refine function features.
- Step 7: End, the conceptual scheme was further designed in detail.

4.2 Knowledge Matching Algorithm

Usually, knowledge is expressed in ordered triples, including design objects, features, and values. E $= (O, c, v)$, Where E is the matter-element, O is the object, c represents the features of the object, and v represents the values of c. An n-dimensional vector of DR between the three (E$_{DR}$) can be expressed as follows [14]:

$$\mathrm{E}_{DR} = (DR, c_{DR}, v_{DR}) = \begin{bmatrix} DR & c_{DR_1} & v_{DR_1} \\ & c_{DR_2} & v_{DR_2} \\ & \cdots & \cdots \\ & c_{DR_n} & v_{DR_n} \end{bmatrix}$$

where c_{DR} is the features of DR, which include quality, function, performance, price, appearance, service life, environmental protection, and other aspects of customer requirements. Personalized demand is also an important part of it. v_{DR} represent the corresponding values of each c_{DR}. Similarly, the functional feature matter-element (E$_{FF}$) can be expressed as:

$$\mathrm{E}_{FF} = (FF, c_{FF}, v_{FF}) = \begin{bmatrix} FF & c_{FF_1} & v_{FF_1} \\ & c_{FF_2} & v_{FF_2} \\ & \cdots & \cdots \\ & c_{FF_n} & v_{FF_n} \end{bmatrix}$$

where c_{FF} is the features of FF, which include material, control, transmission, drive, structure, overall relationship, safety and maintenance performance, etc. v_{FF} represent the corresponding values of each c_{FF}.

$$\mathrm{E}_{FK} = (FK, c_{FK}, v_{FK}) = \begin{bmatrix} FK & c_{FK_1} & v_{FK_1} \\ & c_{FK_2} & v_{FK_2} \\ & \cdots & \cdots \\ & c_{FK_n} & v_{FK_n} \end{bmatrix}$$

In the n-dimensional vector of functional knowledge matter-element (E_{FK}), c_{FK} is the features of FK which include the detailed description, applicable stage, applicable object, type, storage form, domain scope, etc. of the knowledge expression process. v_{FK} represent the corresponding values of each c_{FK}. The features listed in the above three n-dimensional vectors (c_{DR}, c_{FF}, c_{FK}) will be subdivided into several sub-features according to their characteristics, which can be expressed in the form of sub-matter-element.

Based on this, consideration of similarity can achieve matching between different matter elements [10, 12]. If any matter-element D_X, its feature set is C_{D_X}, the matched matter-element feature is D_Y, and its feature set is C_{D_Y}, then the similarity between D_X and D_Y can be calculated.

$Sim(C_{D_X}, C_{D_Y}) = \frac{p}{m+n-p}\sum_{i=1}^{p}(w_i \times S(C_{D_{X_i}}, C_{D_{Y_i}}))$, $S(C_{D_{X_i}}, C_{D_{Y_i}}) \in [0, 1]$ where p represents the number of matter-element that can be used for similarity calculation, and w_i represents the p weight of the ith matter-element. S represents the similarity of the ith matter-element between the existing object D_X and the matching object D_Y. In the mapping process of conceptual design, D_X is generally a clear numerical point, and the D_Y to be matched comes from different ontology databases and has a numerical interval. In this case, Hamming approach degree can be used to calculate $S(C_{D_{X_i}}, C_{D_{Y_i}})$. Assuming that the value interval of $C_{D_{Y_i}}$ is $[v_{D_{Y_{im}}}, v_{D_{Y_{in}}}]$, and the value of $C_{D_{X_i}}$ is $v_{D_{X_i}}$, then the approaching degree between $C_{D_{X_i}}$ and $C_{D_{Y_i}}$ is calculated as follows:

$$S(C_{D_{X_i}}, C_{D_{Y_i}}) = 1 - \frac{\int_{v_{D_{Y_{im}}}}^{v_{D_{Y_{in}}}} |x - v_{D_{X_i}}| dx}{(v_{D_{Y_{in}}} - v_{D_{Y_{im}}})h_i}$$

$$= \begin{cases} 1 - \frac{v_{D_{Y_{im}}} + v_{D_{Y_{in}}} - 2v_{D_{X_i}}}{2h_i} (v_{D_{X_i}} \le v_{D_{Y_{im}}}) \\ 1 - \frac{(v_{D_{Y_{im}}} - v_{D_{X_i}})^2 + (v_{D_{Y_{in}}} - v_{D_{X_i}})^2}{2h_i(v_{D_{Y_{in}}} - v_{D_{Y_{im}}})} (v_{D_{Y_{im}}} \le v_{D_{X_i}} \le v_{D_{Y_{in}}}) \\ 1 - \frac{2v_{D_{X_i}} - v_{D_{Y_{im}}} - v_{D_{Y_{in}}}}{2h_i} (v_{D_{X_i}} \ge v_{D_{Y_{in}}}) \end{cases}$$

where h_i the value range of the matter-element of the matching object, and $h_i = \max(C_{D_{Y_i}}) - \min(C_{D_{Y_i}})$. The two mapping relationships ("requirement-function" mapping, "function-knowledge" mapping) mentioned above can be matched through this similarity calculation. If $Sim(C_{D_X}, C_{D_Y}) > 0$, it means that the match is successful. In the mapping process of ontology-based functional knowledge, while effective knowledge is pushed to the designer, the designer or user can also evaluate and feedback on the pushed knowledge. This helps to upgrade the ontology database and improve the accuracy of knowledge matching. The knowledge or case obtained through knowledge reasoning is essentially an aid and incentive to the conceptual design process, and this process is not limited to knowledge in a single domain. In a sense, knowledge reasoning motivates designers to re-understand and characterize design issues from different perspectives, which is also the key to the application of knowledge in innovative design.

5 Knowledge System for Innovative Conceptual Design

5.1 Overall System Architecture

The knowledge properties and knowledge cases in the knowledge source are fed back to designer through the key information of knowledge system when a search request sent by designer is received by relevant unit module of knowledge system. The overall system architecture of knowledge is proposed as shown in Fig. 5.

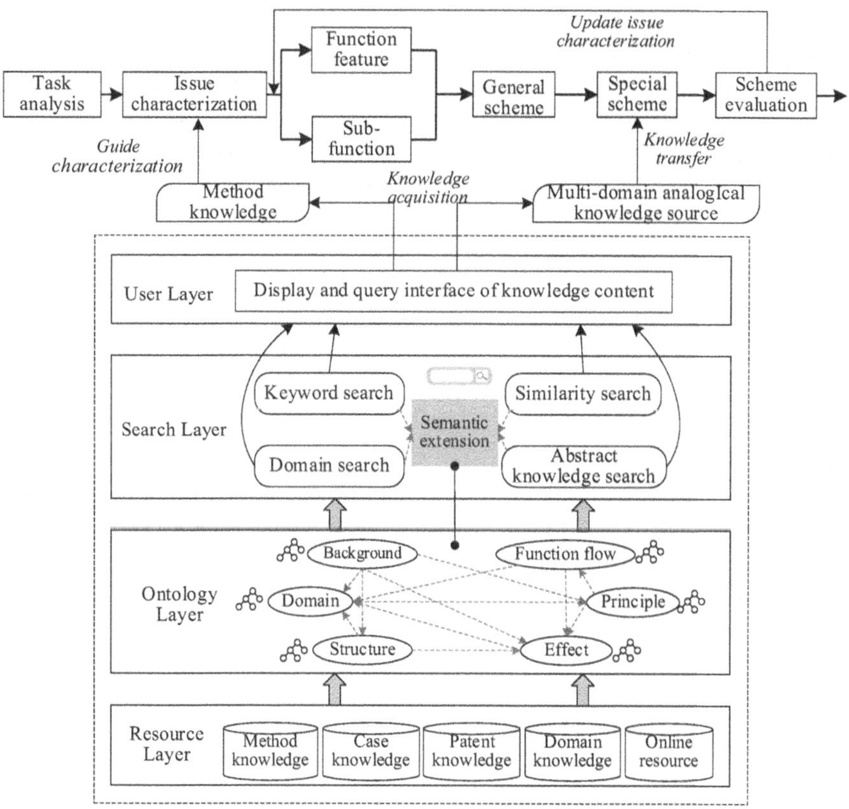

Fig. 5 Overall system architecture of knowledge

There are two parts included in the knowledge-assisted innovative conceptual design: (1) Issues characterization: The classification of design issue is guided or assisted by design knowledge in the knowledge base system when facing design task. It can be expressed as a function or issue-driven starting point. The issues can be characterized by different innovation strategies provided by the system. On the other hand, the issue characterization may be updated by the designer who is inspired under the multi-domain analogical knowledge source search. The issue characterization model is not limited to the "top-down" or "bottom-up" solving process, but the analysis of the issues be expanded from any level of abstraction; (2) Knowledge transfer: Analogical knowledge sources in different fields can be provided by the knowledge system through the search method of the designer. For example, Analogical knowledge with a high degree of relevance in the field can be provided by keyword and similarity search; Analogical knowledge in the near field can be provided by the search of field method; Cross-domain knowledge cases are often obtained through abstract knowledge search. It can solve innovative design issue and carry out cross-domain knowledge analogy transfer by stimulating the designer's imagination.

The overall system architecture of knowledge can be divided into four parts: (1) User layer: Functions such as user login, access, knowledge query, and knowledge display can be realized at this layer; (2) Search layer: Design knowledge related to requirements is provided to design user through different search strategies; (3) Ontology layer: The semantic knowledge expression is formed through the relational network and semantic network between the knowledge attribute of the design resources; (4) Resource layer: Patent knowledge, case knowledge and other knowledge defined by knowledge tags are stored, as is the method knowledge.

5.2 Function Model of Knowledge System

Design knowledge and solving methods can be effectively provided by the knowledge system, and it helps the product design level of the enterprise and the innovative design ability of designer to be improved. Because of various types of knowledge existing in the product design process are effectively integrated and managed by the knowledge system, the knowledge interaction and knowledge support of man–machine system can be realized by pushing the required knowledge to the designer at the appropriate stage to assist him/she in making design decisions. Based on the above, a function model of the knowledge system is proposed as shown in Fig. 6. The user management module, process management module, knowledge management module, and operation management module are included in the function model.

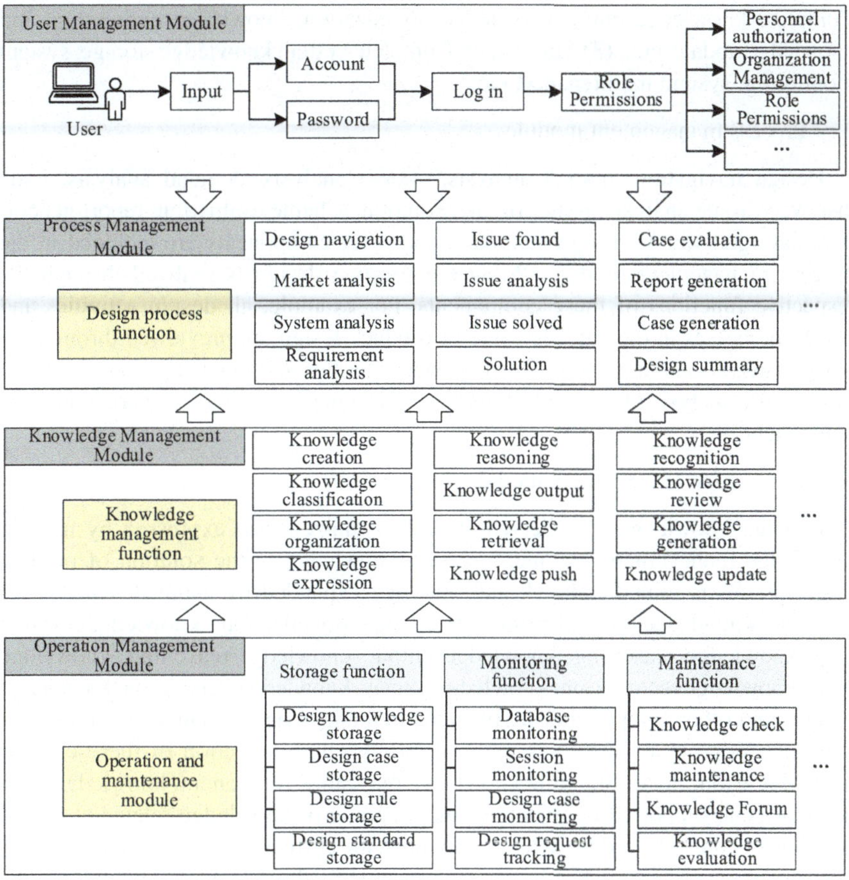

Fig. 6 Function model of knowledge system

1. User management module

User login, role permissions, and other functions are included in the user management module. The account information of user is verified and the legitimacy of user is confirmed through the login function. The role authority function performs role authorization based on account information, and different functional modules can be used by users with different permissions. The role is created according to the product design task, and the corresponding role is set to complete a certain design task. Users are converted into corresponding roles according to their assigned tasks by assigning different users different permissions. There are three kinds of users in this model: (1) Designer: Complete design tasks by acquiring design knowledge and methods, using knowledge reasoning, knowledge push, and other functions; (2) Knowledge manager:

Complete knowledge entry, knowledge organization, knowledge expression, and knowledge update, etc.; (3) Maintainer: Complete system knowledge storage, system monitoring, system maintenance, etc.

2. Process management module

Design navigation, market analysis, system analysis, demand analysis, issue discovery, issue analysis, issue solving, solution, scheme evaluation, report generation, case generation, design summary, and other components are included in the process management module. Product innovative design is realized through the respective functions of these components. For example, all design activities and design processes in the full life cycle of product design are previewed through the design navigation component; Business opportunities and innovations are captured through the analysis of the current market environment acquired by market analysis component.

3. Knowledge management module

Existing knowledge is collected, sorted, organized, and expressed by using a knowledge management module, and new knowledge in the solution of product design issues is mined, refined, and ontology expanded by using it. Knowledge creation, knowledge classification, knowledge organization, knowledge expression, knowledge reasoning, knowledge output, knowledge reasoning, knowledge push, knowledge recognition, knowledge review, knowledge generation, knowledge update, and other components are included in the knowledge management module, and knowledge management is realized through the interaction of these components. For example, through the knowledge reasoning component, knowledge push technology is supported by matching and mapping of knowledge related to design activities; In the knowledge review component, the correctness of the entered knowledge and the new knowledge produced is ensured by the designer's judgment of the knowledge and the knowledge manager's review of the system.

4. Operation management module

The normal operation of the knowledge system is guaranteed by monitoring and maintenance of the operation management module. The storage function module, the monitoring function module, and the maintenance function module are included in the operation management module. Components such as design knowledge storage, design case storage, design rule storage, and design standard storage are included in the storage function module. Components such as database monitoring, session monitoring, design case monitoring, and design request tracking are included in the monitoring function module. Components such as knowledge detection, knowledge maintenance, knowledge forum, and knowledge evaluation are included in the maintenance function module. The normal work of the knowledge system is guaranteed through the joint action of these components. For example, through the knowledge detection component, in order to improve the accuracy of knowledge push, the knowledge mastered by the designer is tested and filtered; The effect of knowledge push is

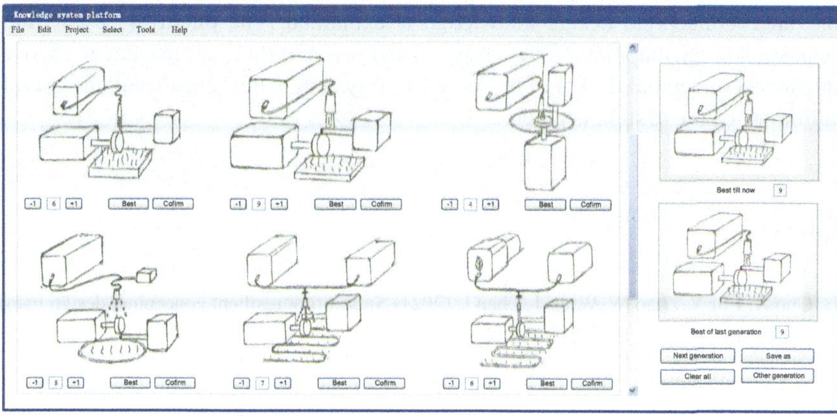

Fig. 7 Interface of knowledge system

evaluated by the knowledge evaluation component. Through this component, whether the designer's issue is solved or how much help is obtained will be judged, thereby providing a theoretical basis for the improvement and upgrade of the knowledge system.

An interface of knowledge system is shown in Fig. 7.

6 Conclusion

Innovative conceptual design has played an important role in stimulating the creativity of designers and assisting in improving the design capabilities of designers. In this chapter, the key design phases where the designer's creativity is concentrated are introduced. The main contents are as follows: (1) Conceptual design is introduced by explaining the importance and characteristics of conceptual design. The main content of innovative conceptual design is introduced, and according to different design goals, the innovative concept design is divided into four types: issue-oriented innovative design, function-oriented innovative design, product-oriented innovative design, and form-oriented innovative design. (2) The main features of the product conceptual design process are introduced, and the design process is divided into four domains for different stages: customer requirements domain, functional structures' domain, conceptual scheme domain, and scheme evaluation domain. (3) An innovative conceptual design process model is presented. The mapping relations between issue space, solution space, and knowledge space in cognitive activities are shown in detail, and between design elements in the space. The mapping between design elements is divided into fourteen mapping processes. (4) The role and methods of knowledge reasoning are introduced. The ontological relation of innovative conceptual design knowledge is explained. The ontology-based functional knowledge mapping process is given and seven of them are presented. (5) The innovative conceptual design of the knowledge system is introduced, and the overall framework of the system is given.

References

1. Guo X, Liu Y, Zhao W, Wang J, Chen L (2021) Supporting resilient conceptual design using functional decomposition and conflict resolution. Adv Eng Inform 48:101262
2. Nomaguchi Y, Fujita K (2013) Knowledge representation framework for interactive capture and management of reflection process in product concepts development. Adv Eng Inform 27:537–554
3. French MJ (1985) Conceptual design for engineers. Design Council, London, New York, Berlin
4. Li Q, Zhao W, Zheng L, Chen L, Zhang K, Guo X (2018) Innovative design of replacement device for vulnerable parts in the nuclear radiation environment. Sci Technol Nucl Installations 2018:1–11
5. Sternberg RJ, Lubart TI (1995) Defying the crowd: cultivating creativity in a culture of conformity. Free Press, New York
6. Li Y, Wang J, Li X, Zhao W (2007) Design creativity in product innovation. Int J Adv Manuf Technol 33:213–222
7. Zheng H, Feng YX, Gao YC (2017) The solving process of conceptual design for complex product based on performance evolution. J Mech Eng 54:214–223
8. Li W, Li Y, Wang J, Liu X (2010) The process model to aid innovation of products conceptual design. Expert Syst Appl 37:3574–3587
9. Jiang T, Nevill GE (2002) Conflict cause identification in web-based concurrent engineering design system. Concurrent Eng Res Appl 10:15–26
10. Guo X, Zhao W, Hu H, Li L, Liu Y, Wang J, Zhang K (2021) A smart knowledge deployment method for the conceptual design of low-carbon products. J Cleaner Prod 321:128994
11. Hatchuel A, Weil B (2008) C-K design theory: an advanced formulation. Res Eng Design 19:181–192
12. Zhang K, Zhao W, Wang J, Chen L, Wang C, Guo X (2016) Research on knowledge support technology for product innovation design based on quality function knowledge deployment. Adv Mech Eng 8:2071834267
13. Horridge M, Jupp S, Moulton G, Rector A, Stevens R, Wroe C (2009) A practical guide to building owl ontologies using protégé 4 and co-ode tools edition 1.2. Univ Manchester 107
14. Zhang K, Zhao W, Wang J, Chen L, Guo X (2019) Knowledge push technology based on quality function knowledge deployment. Proc Inst Mech Eng. Part C, J Mech Eng Sci 233:1119–1138

Establishing Product Appearance Specifications with the Identification of User Aesthetic Needs in Product Conceptual Design

Huicong Hu, Ying Liu, and Wen Feng Lu

This chapter presents the basic concepts of aesthetic needs and product specifications in conceptual design. The user aesthetics needs are considered one of the significant determinants in increasing user satisfaction. In this regard, the importance of establishing product appearance specifications to identify user aesthetic needs is discussed. A method is introduced to demonstrate the significance of considering aesthetic and emotional needs when establishing product appearance specifications in product conceptual design. To improve appearance specifications based on aesthetic experiences collected from users and designers, an approach using fuzzy logic is proposed and illustrated by a case study of digital camera design.

1 Essential of Identifying the Aesthetic Need in Product Conceptual Design

The focus of the conventional conceptual design is typically on functional design, which is targeted at generating appropriate structures so that required functions are provided [5, 56]. The user needs are generally structured into function or

---

H. Hu
Department of Design, School of Humanities and Social Sciences, Harbin Institute of Technology, Shenzhen, Shenzhen 518055, Guangdong, China

Y. Liu (✉)
Department of Mechanical Engineering, School of Engineering, Cardiff University, Cardiff CF24 3AA, UK
e-mail: LiuY81@Cardiff.ac.uk

W. F. Lu
Department of Mechanical Engineering, College of Design and Engineering, National University of Singapore, Singapore 117575, Singapore

P. Kyratsis et al. (eds.), *Advances in Product Design Engineering*, Management and Industrial Engineering,
https://doi.org/10.1007/978-3-030-98124-2_9

usability-related statements, while design specifications are formalised into technical descriptions or values of functions. However, with the improvement of life quality, a major contributing factor contributing to a successful product is user satisfaction [9, 51, 57]. With the aim of meeting the competitive market requirements and enabling the purchase decision of a customer, the functional aspect is not enough. To increase user satisfaction, the product should be able to satisfy user needs to a higher level. In this case, the aesthetic aspect of user needs should be taken into consideration as well [1, 20, 40], especially for electronic consumer products, in which field the trend is getting more obvious.

As the transitional role, which translates the language of the users into technical descriptions of product requirements, product specifications are studied to understand user needs in conceptual design [7]. However, formalised into technical descriptions or values, product specifications are generally associated with usability or functional requirements in most related design studies [8–10]. Additionally, few studies look into the appearance of product specifications, which are characterised by the choices of surface materials and the dimensions of basic geometric. Also, regarding the establishment of product appearance specifications, few studies discussed emotional needs with aesthetic considerations regarding product appearance, and few studies have included the aesthetic aspect of design information.

In terms of quantifying user aesthetic needs, retrieving user responses to the visual appearance of design elements has been discussed as an effective approach to the establishment of product appearance specifications [2]. Under this approach, based on the perceived user reaction and product appearance specifications, the design team is enabled and facilitated to acquire user needs. In a study that investigates the sentimental response of the customer to a product, Kansei Engineering was developed by Nagamachi to translate user psychological feelings into product design elements [41]. By applying this approach, other related studies have been proposed as well [2, 9, 52]. Given the fact that there are successful applications of these methods in processing the emotional needs of users, the studies of Kansei Engineering face a constrain, which is the insufficient considerations of aesthetically attractive visual configurations. These configurations support the arrangement of design elements. Kansei words describe the user needs of product forms, which are usually adjectives, nouns, or verbs that describe specific emotions. As Kansei Engineering explains, it works by connecting Kansei words with design elements. According to existing product structures, it can effectively identify design elements that match Kansei words used by users. It could not, however, specify how the design elements should be arranged. Thus, the newly identified design elements cannot align with the initial product structures in a visually pleasing manner [17].

To consider the aspects of both emotion and the arrangement of design elements, assessing the aesthetic experience of users could be an effective way. The aesthetic experience could reflect how a user perceives and responses to the aesthetic quality of a product regarding both the inherent attractiveness of the arrangement of design elements and the expression of implied emotions and meanings. Therefore, this paper is aimed at proposing a novel approach that supports the identification of user aesthetic needs and the establishment of product specifications in appearance

(appearance specifications). It is aimed to improve appearance specifications by taking both the emotions and the arrangement of design elements into considerations.

First of all, based on the understanding of user aesthetic experience, the proposed approach investigates user aesthetic needs. The way people perceive and respond to a product's aesthetic quality can be indicated by the aesthetic experience, including the emotions and the arrangement of design elements. In this approach, appearance specifications are represented using a comprehensive model. A mapping task is then performed for establishing improved appearance specifications based on aesthetic experience. The mapping task consists of (1) the construction of a mapping model between initial appearance specifications and user aesthetic preferences of user aesthetic experience and (2) the implementation of the mapping model to obtain improved appearance specifications with enhanced user aesthetic preferences. A case study on digital camera designs was conducted to demonstrate the effectiveness of the proposed approach. The proposed method could help to establish appearance specifications-based user aesthetic needs.

2 Background and Related Works

2.1 Aesthetic Information in Conceptual Design

The conceptual design is typically defined as the initial stages of the design process when design solutions are fuzzy [21]. The impact of conceptual design is substantial since a large number of ideas are generated during this process, which is considered to facilitate the achievement of a desirable design [28]. As the conceptual design process converts the design problem at an early stage into a transparent representation of design solutions, the uncertainty and unknown are reduced [35]. With the aim of reducing the uncertainty and the unknown, which lies in the design problem, decomposition of the problem and investigation of potential design solutions are typically conducted before generating a final concept [12]. In this case, the generation of creativity is usually considered to be motivated by the development of design solutions and problems at the same time [14].

The conceptual design process is usually divided into several phases, including user needs identification, specifications establishment of the product, concept generation, selection, and evaluation [55]. First of all, with the aim of identifying user needs, the extraction and organisation of the statements from aimed users will be conducted in hierarchical order. Following the assignment of importance fuzzy, a set of carefully constructed user need statements are generated as the foundation and motivation for design specifications establishment. Subsequently, a rough description of the appearance and structures for the product is created within the product concepts. The information collected along the conceptual design process is essential to product development [42].

Conventionally, the functional and structural aspects are the primary focus in terms of the design information representation [43]. With the aim of delivering the expected effect, the represented information of functional aspects follows geometry restrictions. Meanwhile, the structural aspects focus on the design solutions which deliver the expected effect. For instance, a representation of design knowledge, including structures connections were proposed by Amaresh [6]. Also, with the aim of supporting the functional synthesisation, the function-behaviour-state (FBS) modeller based on knowledge representation [56] was proposed. Based on this FBS modeller, further studies on conceptual design studies were conducted [32, 48].

With the increasing life quality requirement, the main focus in terms of design information is not only on the functional aspect but also on the aesthetic aspect [58]. The definition of aesthetics is the pleasure obtained from sensory [22]. As a fundamental feature for a product, visual product aesthetics include the constitution of the form, colour, and texture of a product preference [4, 28, 59]. Therefore, visual product aesthetics is one of the most essential contributing factors of a desirable product [29].

In the field of design information representation, it is a popular research topic on the consideration of subjective requirements, especially aesthetic ones. With the aim of recognising the aesthetic aspect, it was stated [23] that there are two prominent indicators which are the inherent arrangement of visual design elements and the expression of design emotions implied in its design appearance when aesthetic information is processed. With the aim of realising positive customer satisfaction, aesthetic quality is one of the most contemptuous aspects of product development [39]. In addition, there are high correlations between the purchase decisions made by customers and the visual aesthetics design elements. It is believed that the aesthetic quality can be improved if the aesthetic needs of customers are measured and achieved through product forms design [18]. Aesthetic shapes are characterised by integrity, order, visual balance, rhythm, and appropriate size ratio in aesthetic cognition [15].

The expression of design emotions is one of the essential indicators of aesthetic information [50]. However, the understanding of design emotions expression is subjective since the unquantifiable implications or symbols implied in the product appearance were indicated. In regard to this indicator, Kansei engineering is a popular approach that assesses user-perceived emotions in design and deals with the sentiments of consumers [38]. Kansei engineering was defined as the translating approach of a consumer's psychology regarding the product to the design elements [41]. Kansei engineering has been successfully applied in the design domain as it conveys the emotions in a design. Similar to the principle of Kansei Engineering, the quality function deployment (QFD) transforms the needs of customers into technical requirements [7]. For its applications, the customer preferences are identified using the QFD method in the quality analysis of products [33]. Based on online reviews, Jin et al. focus on engineering characters for QFD in the product design [27].

2.2 Understanding User Aesthetic Needs Through Aesthetic Experience

As a concept, aesthetic experience entails a number of complex processes that are involved in the interaction with a product's visual experience. According to Leder, Belke, Oeberst, and Augustin, aesthetic experience involves perceptual analysis, comparison with previous encounters, classification, interpretation, and evaluative judgments, which eventually result in aesthetic judgments and emotions [37].

An element of visual design can have a significant impact on the aesthetic experience of a product through the expression of information implied in its appearance and its integrated arrangement of elements [23]. These conclusions can also be found in the "classical aesthetics" and "expressive aesthetics" dimensions [10, 36], as well as Crilly's conclusion concerning aesthetic impression, semantic interpretation, and symbolic associations with cognitive response [11].

Depending on the way in which design information is presented, people might understand and view the product differently [23]. Rather than considering tangible artefact characteristics, this indicator takes into account intangible properties of meanings and metaphors that are embedded in the product forms. Due to its subjective nature, this indicator depends greatly on someone's background, identity, personality, social status, or culture, etc. [11, 24]. In addition, the user-perceived attractiveness of a product may also be affected by the typicality and novelty of the product form and by the usage of certain design metaphors or expressions [23, 44].

Form, colour, texture, etc., affect a product's universal appeal as a result of its inherent arrangement in the design. The human mind will award beauty to certain geometric shapes, proportions, and colour combinations [47]. It is an objective property that affects the aesthetic experience and is considered as reflecting the inherent attractiveness that is perceived by sensors. As a consequence of the notion of design beauty being an important part of many design theories, such as the golden section [13], many have been put forward. The Bauhaus is one of the famous pioneers of the field of product design. The Bauhaus' teaching theory is often embraced in product design because it actually incorporates Gestalt psychology and tries to create a "new sense" from the design elements like line, colour, text, and so on [31]. Those who follow Gestalt psychology—which was developed in the early 1900s rather than the modern generation-believe that "there is more to an experience of the whole than its parts" [30]. As a result, people tend to perceive things that are pleasing, balanced, and unified, spreading a feeling of overall harmony [8]. In spite of this, a product considered to be too harmonious may be regarded as boring and monotonous [23]. The ability of complex and varied experiences to generate arousal can sometimes be required in some circumstances [3]. There are a number of aesthetic equilibrium theorists who suggest that this balance lies between boredom and confusion [16].

According to aesthetics, design elements are vocabularies that constitute a design form [31]. There are several commonly recognised design elements, including line, shape, colour, and texture [53]. A number of researchers have argued that an appealing psychological form is one that holds the right aesthetic balance between covenant order and complex arousal based on research findings. It is critical to apply aesthetic principles to provide a heuristic guide for users in order to be able to perceive both arousal-reducing and arousal-driving design strategies [31, 34]. The principles of aesthetic design can be defined as universally recognised compositional strategies for visual appearance. In addition to constituting and arranging the elements of visual design, they contribute to the aesthetics of the process [19]. Based on contrast, rhythm, balance, and proportion (CRBP), Stebbing enumerated the basic principles of aesthetic design on visual composition [54]. A contrast is, in this context, the juxtaposition of different elements of a composition that creates visual disunity. The concept of rhythm is based on repeated or alternative elements that are arranged with organised or defined intervals. With the aim of creating a sense of equilibrium, balance is the application of design elements. Various elements in a design must be sized and scaled so that they are proportionately large [25].

3 Overall Framework

As appearance specifications are usually described as semantic requirements, which are imprecise and challenging to quantify, design specifications in conceptual design are typically composed of a collection of attributes or metrics with certain values. To describe the requirements of product appearances, a representation model of appearance specifications is first proposed in order to recognise important attributes.

As illustrated in Fig. 1, a representation model for appearance specifications is presented. It is useful to use appearance specifications to describe the appearance of the product from the six different points of view (front, back, top, bottom, left, and right). Within each view, two categories of appearance specification are displayed, namely emotion and aesthetic indicator. Emotions, or content meanings implied in the appearance of the product, are defined as the attributes of subjective emotions and meanings. Various emotional words are included in this product. A noun may refer to an object, or an adjective might describe certain psychological meanings. An emotional adjective is employed to evaluate the expression of design forms in this study. As stated by Dong et al. [46], four categories are used to categorise adjectival words, these categories being physiological, psychological, cultural, and physical. The adjectives in each category are each considered attributes of emotion corresponding to implied feelings associated with the product's appearance. When defining appearance specifications, the value of emotion attribute is determined as the degree to which the product appearance gives people a sense of the emotion word. In the case of the "modern" emotion attribute, values could be "very high", "high", "middle", "low", and "very low". It indicates that the product may not appear modern to people if the value of the emotion attribute "modern" is "very low = 1". It is also possible to refer to an emotion attribute as a noun, which could refer to the shape of an actual object. This could be an inanimate object like a cloud or an animate object like a beetle. An abstract notion, such as Sony or Modernism, can also be a representation of an idea. As described in the aesthetic indicator, design

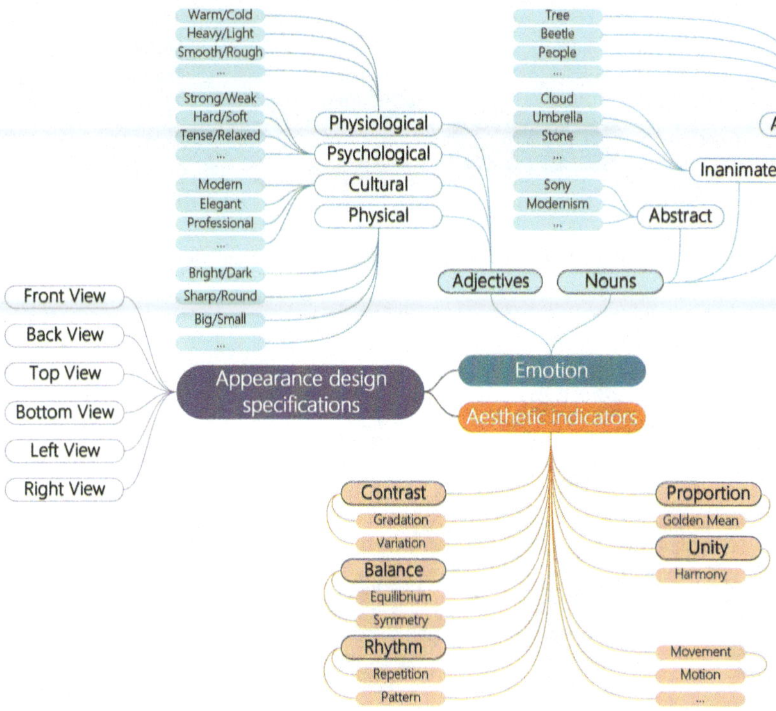

Fig. 1 Representation of appearance specifications

elements must be arranged according to specific strategies. The aesthetics of design principles determine how to arrange design elements. Aesthetic design principles are well-known and are used by designers in composing design elements, as discussed in previous chapters. They are regarded as attributes of aesthetic indicators, as described in previous chapters. According to Stebbing's study [54], the representation model contains common aesthetic design principles, such as contrast, balance, rhythm, proportion, and unity. As a result, the aesthetic indicator attribute value is determined by the degree of implementation of the corresponding aesthetic design principle in the appearance of the product.

As shown in Fig. 2, the front view of a digital camera is represented by appearance specifications. Aesthetic design principles are contrast and balance as the elements of the aesthetic design indicator, and emotion adjectives "classic", "modern", "elegant", and "cute" are selected as emotion attributes.

Next, the overall method for improving appearance specifications is presented in Fig. 3. Based on the aesthetic experience of existing design samples, a mapping task is performed to generate appearance specifications. The mapping task consists of three steps. The first step is to acquire initial appearance specifications and user aesthetic preferences of existing design samples. The values of initial appearance specifications are obtained from both evaluations of user aesthetic experience and

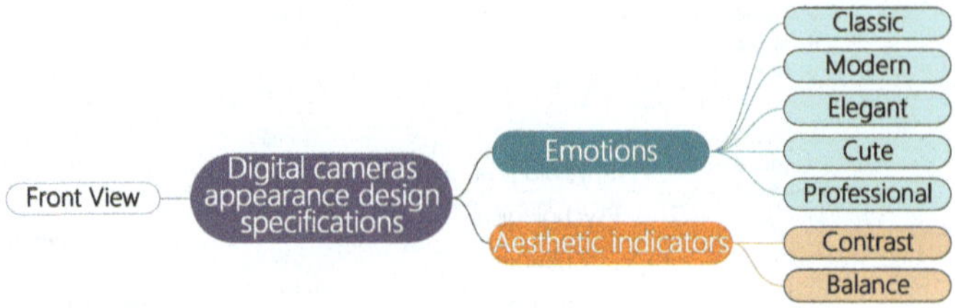

Fig. 2 Representation of digital camera appearance specifications (front view)

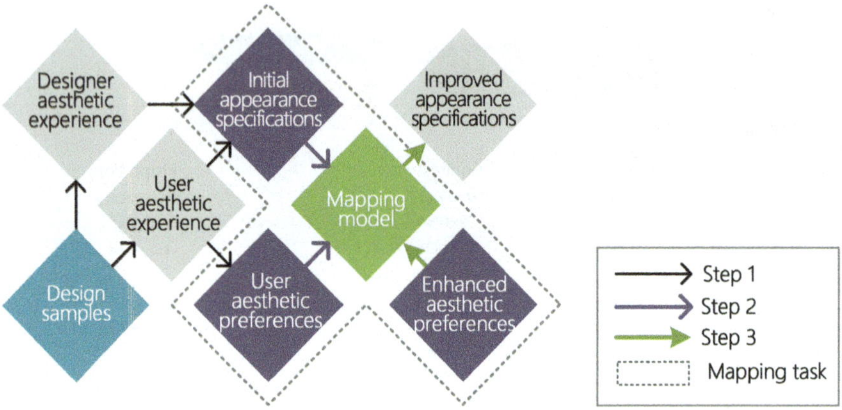

Fig. 3 The proposed method for improving appearance specifications based on aesthetic experience

designer aesthetic experience. In the evaluation of user aesthetic experience, the user would also indicate their aesthetic preferences of the existing design samples. Step 2 focuses on constructing a mapping model between the initial appearance specifications and the user aesthetic preferences. Step 3 is to generate improved appearance specifications based on enhanced user aesthetic preferences, which are predefined by designers.

Figure 4 shows the detailed process of constructing the mapping model between initial appearance specifications and user aesthetic preferences (Step 1 and Step 2). In Step 1, design samples are chosen from existing product designs or prototypes. From the design samples, the attributes of the initial appearance specification are determined based on the proposed representation model. With the aim of obtaining the values of the initial appearance specification attribute regarding design samples, user evaluation and designer evaluation are conducted. Participants of the user evaluation could be users from the target user group with defined personal backgrounds. Participants of the designer evaluation are designers with certain design experience and knowledge of applying aesthetic design principles. During the evaluation, users

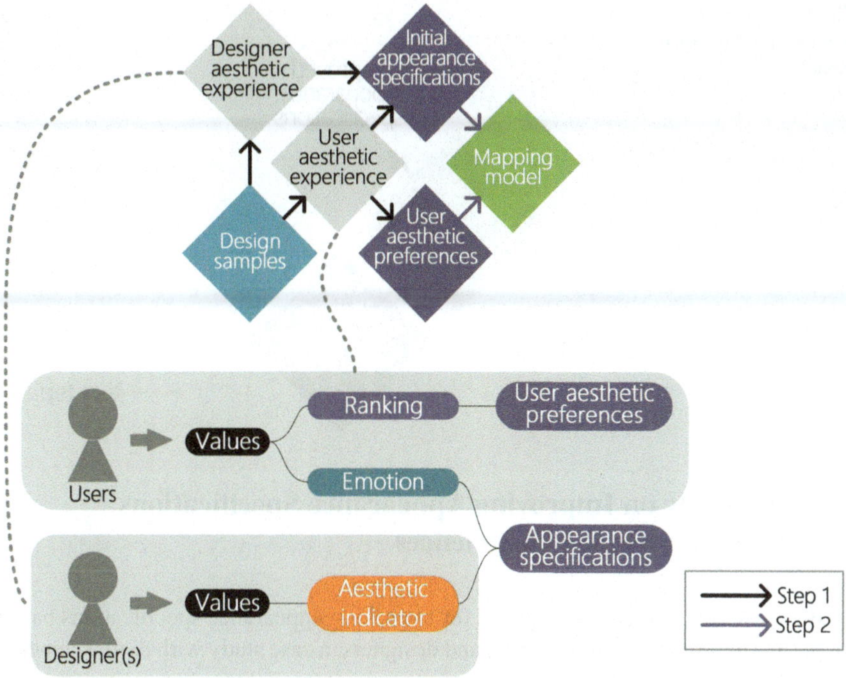

Fig. 4 Process of constructing the mapping model

and designer(s) observe the appearance of design samples and gain certain aesthetic experiences. Their aesthetic experiences could be acquired from survey questions regarding each attribute of initial appearance specifications. In the user evaluation, values of the emotion attributes of initial appearance specifications and ranking values of user aesthetic preferences are obtained. In the designer evaluation, designer(s) evaluate the appearance of design samples and indicate the values of aesthetic indicator attributes of initial appearance specifications. The resulted values of both user and designer evaluations would be used for constructing the mapping model. In Step 2, the mapping model is constructed with the input of initial appearance specifications and the output of user aesthetic preferences.

Step 3 is to implement the constructed mapping model to generate improved appearance specifications. Figure 5 shows the detailed process of implementing the mapping model for the generation of improved appearance specifications. In this process, enhanced user aesthetic preferences are first defined as the target aesthetic preferences. Based on the mapping model, improved appearance specifications that result in the target aesthetic preferences could be generated.

Fig. 5 Process of implementing the mapping model

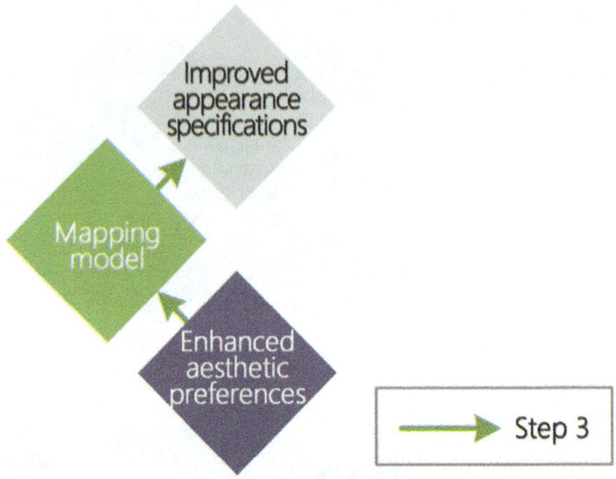

4 Case Study on Improving Appearance Specifications Based on Aesthetic Experiences

To demonstrate the method proposed for improving appearance specifications based on the aesthetic experiences of users and designers, a case study with digital cameras was conducted. By applying fuzzy logic, a mapping model was constructed for this study. There is evidence to suggest that fuzzy logic is an excellent tool for modelling information with imprecise values that depend on their degree [45]. Therefore, it can be used to construct mappings between original appearance design specifications and aesthetic preferences. Below is an outline of the steps involved in this case study.

1. Select design samples and attributes of initial appearance specifications for aesthetic experience evaluation.
2. Determine linguistic variables for initial appearance specifications.
3. Construct a fuzzy set X of initial appearance specifications.
4. Determine linguistic variables for user aesthetic preferences and construct a fuzzy set Y of user aesthetic preferences.
5. Construct fuzzy rules between initial appearance specification fuzzy set X and user aesthetic preference fuzzy set Y

   Rule: IF X_1 is A_1 AND X_2 is A_2 ... AND X_n is A_n THEN Y is B,

   where A_1, A_2, ..., A_n and B are fuzzy linguistic values, taken by the input linguistic variables X_i and the output linguistic variable Y.
6. Generate improved appearance specifications with enhanced user aesthetic preferences.

This case study is to demonstrate the appearance design of the interchangeable-lens digital camera. The goal was to establish appearance specifications of the digital camera for the target users of 20–30 years olds female college students.

4.1 *Acquiring Initial Appearance Specifications and User Aesthetic Preferences*

A total of eight interchangeable-lens digital cameras were chosen as the study's design samples. There were different combinations of styling forms, colours, and textures used to select the design samples. Aesthetic design principles "contrast" and "balance" were chosen as aesthetic indicator attributes, whereas emotion attributes "classic", "modern", "elegant", "cute", and "professional" were chosen as emotion attributes.

To determine linguistic variables for initial appearance specifications, the triangular form of the membership function, which is frequently used for representing fuzzy numbers and significantly simplifies the modelling process [26], was employed to determine approximate interval values of linguistic variables. Seven linguistic variables were chosen as criteria for appearance specification attributes (Fig. 6, Table 1).

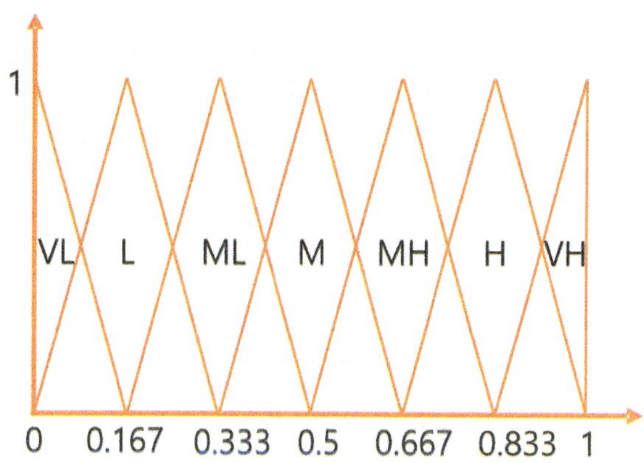

Fig. 6 Membership functions for appearance specification attributes

Table 1 Linguistic criteria for appearance specification attributes

Linguistic variable	Interval of triangular fuzzy number
Very low (VL)	[0, 0.167]
Low (L)	[0, 0.333]
Medium low (ML)	[0.167, 0.5]
Medium (M)	[0.333, 0.667]
Medium high (MH)	[0.5, 0.833]
High (H)	[0.667, 1]
Very high (VH)	[0.833, 1]

Next, the value of each attribute of appearance specifications was obtained. To construct a fuzzy set on emotion attributes, the semantic differential (SD) method, which is a self-report method using a Likert scale, was employed to evaluate design samples regarding each emotion attribute. A seven-point SD scale (1–7) corresponding to the seven linguistic variables (VL to VH) was employed in the user evaluation. Forty-three female college students were selected from target users for the user evaluation. The evaluation result of each emotion attribute is shown in Table 2, with the average and standard deviation values. The converted values of fuzzy sets on emotion attributes are listed in Table 3.

The Analytic Hierarchy Process (AHP) developed by Saaty [49] was applied to perform comparative judgments to determine the degree to which aesthetic principles were implemented in particular design samples in order to construct a fuzzy set of aesthetic indicators. Table 4 illustrates the pairwise comparisons made between design samples in regards to aesthetic design principles based on their intensity of importance, as shown in the figure. By applying local weights, each aesthetic design

Table 2 Users SD evaluation results

Design sample	Classic		Modern		Elegant		Cute		Professional	
	Avg	Std	Avg	Std	Avg	Std	Avg	Std	Avg	Std
1	2.21	1.29	5.95	0.99	5.58	1.10	4.53	1.53	4.23	1.74
2	5.91	1.03	2.56	1.33	4.40	1.50	1.98	0.90	5.14	1.39
3	2.05	0.89	5.65	1.27	3.12	1.67	5.70	1.50	3.28	1.60
4	2.91	1.51	4.56	1.56	3.07	1.59	4.30	1.59	3.42	1.50
5	5.28	1.83	3.42	1.73	4.51	1.53	2.26	1.08	5.84	1.03
6	3.81	1.79	3.23	1.87	3.05	1.49	3.21	2.03	3.98	1.80
7	3.26	1.73	4.67	1.43	4.37	1.75	3.16	1.41	4.70	1.46
8	2.88	1.50	4.88	1.83	6.21	1.05	4.16	1.74	2.95	1.70

Table 3 Values of the fuzzy set on emotion attributes

Design sample	Classic	Modern	Elegant	Cute	Professional
1	L	H	H	MH	M
2	H	ML	M	L	MH
3	L	H	ML	H	ML
4	ML	MH	ML	M	ML
5	MH	ML	MH	L	H
6	M	ML	ML	ML	M
7	ML	MH	M	ML	MH
8	ML	MH	H	M	ML

Table 4 The intensity of importance scale for AHP evaluation

Intensity of importance	Definition	Explanation
1	Equal importance	Two activities contribute equally to the objective
3	Weak importance of one over another	Experience and judgement slightly favour one activity over another
5	Essential or strong importance	Experience and judgement strongly favour one activity over another
7	Demonstrated importance	Activity is strongly favoured, and its dominance demonstrated in practice
9	Absolute importance	The evidence favouring one activity over another is of the highest possible order of affirmation
2, 4, 6, 8	Intermediate values between the two adjacent judgments	When compromise is needed

Table 5 AHP results of the fuzzy set on aesthetic indicator attributes

Design sample	Contrast				Balance			
	Weight	weight *0.917	Linguistic variable \DOS		Weight	weight *0.833	Linguistic variable \DOS	
1	0.126	0.116	L\0.70	VL\0.30	0.848	0.707	MH\0.76	H\0.24
2	0.082	0.075	VL\0.55	L\0.45	0.196	0.163	L\0.98	VL\0.02
3	1.000	0.917	H\0.50	VH\0.50	0.722	0.602	MH\0.61	M\0.39
4	0.746	0.684	MH\0.90	H\0.10	0.428	0.357	ML\0.86	M\0.14
5	0.280	0.257	ML\0.54	L\0.46	1.000	0.833	H\1	VH\0
6	0.530	0.486	M\0.92	ML\0.08	0.277	0.231	L\0.61	ML\0.39
7	0.076	0.070	VL\0.58	L\0.42	0.361	0.301	ML\0.81	L\0.19
8	0.113	0.104	L\0.62	VL\0.38	0.236	0.197	L\0.82	ML\0.18
	Inconsistency = 0.04				Inconsistency = 0.07			

principle was assessed in relation to its degree of implementation. This process was performed by an aesthetic designer with over five years of experience in implementing aesthetic design principles. The resulting AHP weights are presented in Table 5. The largest weights of "contrast" and "balance" were found in Sample 3 and Sample 5, respectively. The linguistic variables regarding "contrast" for Sample 3 and "balance" for Sample 5 were determined as "VH" and "H", respectively by the designer. The corresponding numerical values of the linguistic variables "VH" and "H" were 0.917 and 0.833, respectively, based on the defined linguistic criteria

Table 6 Linguistic criteria for user aesthetic preferences

Linguistic variable	Interval of triangular fuzzy number
Rank 1 (R1)	[0, 0.143]
Rank 2 (R2)	[0, 0.286]
Rank 3 (R3)	[0.143, 0.429]
Rank 4 (R4)	[0.286, 0.571]
Rank 5 (R5)	[0.429, 0.714]
Rank 6 (R6)	[0.571, 0.857]
Rank 7 (R7)	[0.714, 1]
Rank 8 (R8)	[0.875, 1]

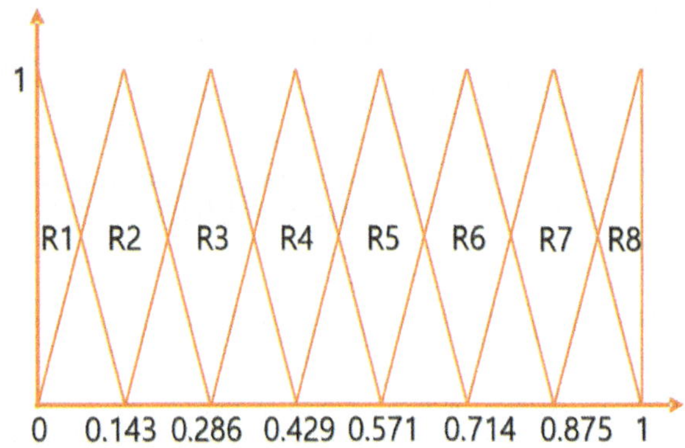

Fig. 7 Membership functions for user aesthetic preferences

of appearance specification attributes (Fig. 6, Table 1). By multiplying the resulting AHP "contrast" weights by the "Contrast" numerical values (0.917) of Sample 3, the "contrast" numerical values of other samples were obtained. The "balance" numerical values of other samples were generated in the same way. The numerical values were then converted into the fuzzy set, which consists of linguistic variables with the degree of support (DOS) based on the defined linguistic criteria of appearance specification attributes. The DOS could be regarded as the weight of the corresponding fuzzy rule for constructing the fuzzy model. Table 5 shows the AHP evaluation results of the fuzzy set on aesthetic indicator attributes "contrast" and "balance".

With the aim of acquiring user aesthetic preferences, user ranking was used to represent the user aesthetic preferences. Table 6 shows the evaluation criteria for linguistic judgments of user aesthetic preferences and Fig. 7 presents the membership functions for user aesthetic preferences. Table 7 shows the ranking results of the fuzzy set on user aesthetic preferences.

Table 7 Ranking results of the fuzzy set on user aesthetic preferences

Design sample	User ranking
1	Ranking 1 (R1)
2	Ranking 5 (R5)
3	Ranking 4 (R4)
4	Ranking 7 (R7)
5	Ranking 3 (R3)
6	Ranking 8 (R8)
7	Ranking 6 (R6)
8	Ranking 2 (R2)

4.2 Constructing the Fuzzy Model

Based on fuzzy sets of the appearance specifications and aesthetic preferences, 32 fuzzy rules were constructed, as shown in Table 8. The weight of each fuzzy rule was determined by the DOS of the aesthetic indicator of "contrast" and "balance".

4.3 Establishing Improved Appearance Specifications

The enhanced user aesthetic preferences are predefined by designers to establish improved appearance specifications based on the constructed fuzzy model. In this case study, referring to the linguistic criteria of user aesthetic preferences, the fuzzy linguistic value of enhanced user aesthetic preferences was defined as R1 (Ranking 1) [0, 0.143]. Based on the constructed fuzzy model, every possible combination of appearance specifications that contributes to the enhanced user aesthetic preferences were searched and generated as improved appearance specifications. Twelve groups of improved appearance specifications that result in Ranking 1 of user aesthetic preferences were generated and are shown in Table 9. The results suggested that users would prefer designs with a lower degree of "contrast" and higher degrees of "modern" and "elegant". The values of appearance specification attributes could provide directions and insights in both aspects of emotions and the arrangement of design elements for designers to generate design concepts.

5 Conclusions

When it comes to increasing user satisfaction, the aesthetic aspect of product design becomes crucial. Conceptual design literature rarely addresses the information representation of appearance specifications. An information representation model of appearance specifications was proposed to address this issue. Further, a method

Table 8 Fuzzy rules between appearance specifications and user aesthetic preferences

	Expression					Arrangement		Weight	User preference
	Classic	Modern	Elegant	Cute	Professional	Contrast	Balance		Ranking
1	L	H	H	MH	M	L	MH	0.5320	R1
2	L	H	H	MH	M	L	H	0.1680	R1
3	L	H	H	MH	M	VL	MH	0.2280	R1
4	L	H	H	MH	M	VL	H	0.0720	R1
5	H	ML	M	L	MH	VL	L	0.5390	R5
6	H	ML	M	L	MH	VL	VL	0.0110	R5
7	H	ML	M	L	MH	L	L	0.4410	R5
8	H	ML	M	L	MH	L	VL	0.0090	R5
9	L	H	ML	H	ML	H	MH	0.3050	R4
10	L	H	ML	H	ML	H	M	0.1950	R4
11	L	H	ML	H	ML	VH	MH	0.3050	R4
12	L	H	ML	H	ML	VH	M	0.1950	R4
13	ML	MH	ML	M	ML	MH	ML	0.7740	R7
14	ML	MH	ML	M	ML	MH	M	0.1260	R7
15	ML	MH	ML	M	ML	H	ML	0.0860	R7
16	ML	MH	ML	M	ML	H	M	0.0140	R7
17	MH	ML	MH	L	H	ML	H	0.5400	R3
18	MH	ML	MH	L	H	ML	VH	0.0000	R3
19	MH	ML	MH	L	H	L	H	0.4600	R3
20	MH	ML	MH	L	H	L	VH	0.0000	R3
21	M	ML	ML	ML	M	M	L	0.5612	R8
22	M	ML	ML	ML	M	M	ML	0.3588	R8
23	M	ML	ML	ML	M	ML	L	0.0488	R8
24	M	ML	ML	ML	M	ML	ML	0.0312	R8
25	ML	MH	M	ML	MH	VL	ML	0.4698	R6
26	ML	MH	M	ML	MH	VL	L	0.1102	R6
27	ML	MH	M	ML	MH	L	ML	0.3402	R6
28	ML	MH	M	ML	MH	L	L	0.0798	R6
29	ML	MH	H	M	ML	L	L	0.5084	R2
30	ML	MH	H	M	ML	L	ML	0.1116	R2
31	ML	MH	H	M	ML	VL	L	0.0684	R2
32	ML	MH	H	M	ML	VL	ML	0.3116	R2

Table 9 Generated improved appearance specifications

Group	Contrast	Balance	Classic	Modern	Elegant	Cute	Professional
1	VL	MH	L	H	H	MH	ML
2	VL	MH	L	H	H	MH	M
3	VL	MH	L	H	H	MH	MH
4	VL	H	L	H	H	MH	ML
5	VL	H	L	H	H	MH	M
6	VL	H	L	H	H	MH	MH
7	L	MH	L	H	H	MH	ML
8	L	MH	L	H	H	MH	M
9	L	MH	L	H	H	MH	MH
10	L	H	L	H	H	MH	ML
11	L	H	L	H	H	MH	M
12	L	H	L	H	H	MH	MH

that supports designers in improving appearance specifications based on aesthetic experience was developed. In the method, a mapping model was constructed with the input of initial appearance specifications and the output of user aesthetic preferences. From the mapping model, improved appearance specifications that result in enhanced user aesthetic preferences could be obtained. A digital camera design case study was conducted to illustrate the overall method. The case study reflected that the proposed method is effective in improving appearance specifications. It also showed that fuzzy logic is applicable to construct the mappings between appearance specifications and user aesthetic preferences.

References

1. Barnes C, Lillford SP (2009) Decision support for the design of affective products. J Eng Des 20:477–492
2. Barone S, Lombardo A, Tarantino P (2007) A weighted logistic regression for conjoint analysis and Kansei engineering. Qual Reliab Eng Int 23:689–706
3. Berlyne DE (1966) Curiosity and exploration. Science 153:25–33
4. Bloch PH (1995) Seeking the ideal form: product design and consumer response. J Mark 59:16–29
5. Chakrabarti A, Bligh TP (1996) An approach to functional synthesis of mechanical design concepts: theory, applications, and emerging research issues. AI EDAM 10:313–331
6. Chakrabarti A, Bligh TP (1994) An approach to functional synthesis of solutions in mechanical conceptual design. Part I: introduction and knowledge representation. Res Eng Design 6:127–141
7. Chan L-K, Wu M-L (2002) Quality function deployment: a literature review. Eur J Oper Res 143:463–497

8. Chang D, Nesbitt KV (2006) Identifying commonly-used gestalt principles as a design framework for multi-sensory displays. In: 2006 IEEE International conference on systems, man and cybernetics. IEEE, pp 2452–2457
9. Chen C-C, Chuang M-C (2008) Integrating the Kano model into a robust design approach to enhance customer satisfaction with product design. Int J Prod Econ 114:667–681
10. Coates D (2003) Watches tell more than time: product design, information, and the quest for elegance. McGraw-Hill Companies, London
11. Crilly N, Moultrie J, Clarkson PJ (2004) Seeing things: consumer response to the visual domain in product design. Des Stud 25:547–577
12. Daalhuizen J, Cash P (2021) Method content theory: towards a new understanding of methods in design. Des Stud 75:101018
13. Elam K (2001) Geometry of design: studies in proportion and composition. Princeton Architectural Press, New York
14. Gero JS, Milovanovic J (2020) A framework for studying design thinking through measuring designers' minds, bodies and brains. Des Sci 6:1–40
15. Göhlich D, Bender B, Fay T-A et al (2021) Product requirements specification process in product development. Proc Des Soc 1:2459–2470
16. Gombrich E (1984) The sense of order: a study in the psychology of decorative art, 2nd edn. Phaidon, Oxford
17. Guo F, Qu Q-X, Nagamachi M et al (2020) A proposal of the event-related potential method to effectively identify kansei words for assessing product design features in Kansei engineering research. Int J Ind Ergon 76:102940
18. Han J, Forbes H, Schaefer D (2021) An exploration of how creativity, functionality, and aesthetics are related in design. Res Eng Design 32:289–307
19. Han J, Sarica S, Shi F et al (2021) Semantic networks for engineering design: a survey. Proc Des Soc 1:2621–2630
20. Hashmi HBA, Shu C, Haider SW et al (2021) Bridging the gap between product design and customer engagement: role of self-determined needs satisfaction. SAGE Open 11:21582440211056600
21. Hay L, Duffy AH, Mcteague C et al (2017) A systematic review of protocol studies on conceptual design cognition: design as search and exploration. Des Sci 3:1–36
22. Hekkert P (2006) Design aesthetics: principles of pleasure in design. Psychol Sci 48:157
23. Hekkert P, Schifferstein HN (2008) Introducing product experience. In: Product experience. Elsevier, Amsterdam, pp 1–8
24. Holbrook MB, Schindler RM (1994) Age, sex, and attitude toward the past as predictors of consumers' aesthetic tastes for cultural products. J Mark Res 31:412–422
25. Huicong H, Wen-Feng L (2020) Design specification representation for intelligent product appearance design. In: E3S Web of conferences, p 02004
26. Igwe PC (2008) Deformable VSOFM and physics-based modeling for conceptual design. Faculty of Graduate Studies, University of Western Ontario
27. Jin J, Ji P, Liu Y (2014) Prioritising engineering characteristics based on customer online reviews for quality function deployment. J Eng Des 25:303–324
28. Jing L, Li Z, Peng X et al (2019) A relative equilibrium decision approach for concept design through fuzzy cooperative game theory. J Comput Inf Sci Eng 19:1–24
29. Kang X (2020) Aesthetic product design combining with rough set theory and fuzzy quality function deployment. J Intell Fuzzy Syst 39:1131–1146
30. Katz D (1950) Gestalt psychology: its nature and significance. The Ronald Press, New York
31. Kim N (2006) A history of design theory in art education. J Aesthetic Educ 40:12–28
32. Komoto H, Tomiyama T (2012) A framework for computer-aided conceptual design and its application to system architecting of mechatronics products. Comput Aided Des 44:931–946
33. Kowalska M, Pazdzior M, Krzton-Maziopa A (2018) Implementation of QFD method in quality analysis of confectionery products. J Intell Manuf 29:439–447
34. Kumar M, Garg N (2010) Aesthetic principles and cognitive emotion appraisals: how much of the beauty lies in the eye of the beholder? J Consum Psychol 20:485–494

35. Lasso S, Kreye M, Daalhuizen J et al (2020) Exploring the link between uncertainty and project activities in new product development. J Eng Des 31:531–551
36. Lavie T, Tractinsky N (2004) Assessing dimensions of perceived visual aesthetics of web sites. Int J Hum Comput Stud 60:269–298
37. Leder H, Belke B, Oeberst A et al (2004) A model of aesthetic appreciation and aesthetic judgments. Br J Psychol 95:489–508
38. Liu S, Stebner AP, Kappes BB et al (2021) Machine learning for knowledge transfer across multiple metals additive manufacturing printers. Add Manuf 39:101877
39. Ma B, Hauer RJ, Xu C (2020) Effects of design proportion and distribution of color in urban and suburban green space planning to visual aesthetics quality. Forests 11:278
40. Mcdonagh D, Bruseberg A, Haslam C (2002) Visual product evaluation: exploring users' emotional relationships with products. Appl Ergon 33:231–240
41. Nagamachi M (1995) Kansei engineering: a new ergonomic consumer-oriented technology for product development. Int J Ind Ergon 15:3–11
42. Nakata C, Hwang J (2020) Design thinking for innovation: composition, consequence, and contingency. J Bus Res 118:117–128
43. Narsale S, Chen Y, Mohan M et al (2019) Design ideator: a conceptual design toolbox. J Comput Inf Sci Eng 19:041007
44. Noble CH, Kumar M (2010) Exploring the appeal of product design: a grounded, value-based model of key design elements and relationships. J Prod Innov Manag 27:640–657
45. Patyra MJ, Mlynek DJ (2012) Fuzzy logic: implementation and applications. Springer Science & Business Media, Berlin
46. Raffaï RL, Dong L-M, Farese RV et al (2001) Introduction of human apolipoprotein E4 "domain interaction" into mouse apolipoprotein E. Proc Natl Acad Sci 98:11587–11591
47. Ralls K (1995) But is it science? Conserv Biol 9:983–984
48. Roy U, Pramanik N, Sudarsan R et al (2001) Function-to-form mapping: model, representation and applications in design synthesis. Comput Aided Des 33:699–719
49. Saaty TL (1988) What is the analytic hierarchy process? In: Mathematical models for decision support. Springer, pp 109–121
50. Saraiva M, Ayanoğlu H (2019) Emotions and emotions in design. In: Emotional design in human-robot interaction. Springer, pp 57–70
51. Smith GC, Smith S (2012) Latent semantic engineering—a new conceptual user-centered design approach. Adv Eng Inform 26:456–473
52. Smith S, Smith GC, Chen Y-R (2013) A KE-LSA approach for user-centered design. J Intell Manuf 24:919–933
53. Spratt F (1987) Art production in discipline based art education. J Aesthetic Educ 21:197–204
54. Stebbing PD (2004) A universal grammar for visual composition? Leonardo 37:63–70
55. Ulrich KT, Eppinger SD (2012) Product design and development. McGraw Hill, London
56. Umeda Y, Ishii M, Yoshioka M et al (1996) Supporting conceptual design based on the function-behavior-state modeler. Ai Edam 10:275–288
57. Wang T, Zhou M (2020) A method for product form design of integrating interactive genetic algorithm with the interval hesitation time and user satisfaction. Int J Ind Ergon 76:102901
58. Wang Y-X (2011) Effects of feature parameters on the form-image sensation of products. J Eng Des 22:181–199
59. Yamamoto M, Lambert DR (1994) The impact of product aesthetics on the evaluation of industrial products. J Prod Innov Manag 11:309–324

Digital Transformation of the Product Design and Idea Generation Process

Gojko Vladić, Nemanja Kašiković, Saša Petrović, Gordana Bošnjaković, and Bojan Banjanin

This chapter presents the transformation of the product design and idea generation process influenced by modern digital technologies. Influences on the first stages of the product design process are considered. Market research as the information source is subject to drastic changes due to the introduction of data gathering techniques in the digital era and emerging of the product intelligence process and tools. The introduction of artificial intelligence in concept design and idea generation offers exciting possibilities in creative design problem solving, suggesting future automation of the design process. Furthermore, the cost-effectiveness and speed of the design development stage are improving through the adoption of virtual and augmented reality alongside rapid prototyping technologies.

G. Vladić (✉) · N. Kašiković · S. Petrović · G. Bošnjaković · B. Banjanin
Department of Graphic Engineering and Design, Faculty of Technical Sciences, University of Novi Sad, Trg Dositeja Obradovića 6, 21000 Novi Sad, Serbia
e-mail: vladicg@uns.ac.rs

N. Kašiković
e-mail: knemanja@uns.ac.rs

S. Petrović
e-mail: petrovic.sasa@uns.ac.rs

G. Bošnjaković
e-mail: gordana.delic@uns.ac.rs

B. Banjanin
e-mail: bojanb@uns.ac.rs

P. Kyratsis et al. (eds.), *Advances in Product Design Engineering*, Management and Industrial Engineering,
https://doi.org/10.1007/978-3-030-98124-2_10

1 Research, Idea Generation, and Concept Design in the Product Design Process

Product development involves a range of activities from the conceptualization to the realization of a new product, making it complex with the need to involve numerous fields of expertise. Usually, it follows a protocol consisting of several stages: idea generation (Ideation), product definition, prototyping, detailed design, validation/testing, and commercialization.

The first stage, idea generation (Ideation), might be the most important as this is the stage where product concepts originate. Making the wrong product concept wastes time and other resources without the chance of achieving the profit objectives. Shocking estimation is that roughly 40% of new products fail at the launching stage, and industry reports that only 13% of new product investments achieve their profit objectives [1, 2].

The idea generation process is highly connected and associated with creativity, invention, and technical innovation. Design solution does not come from a vacuum and the designer's inspiration only. New product ideas might originate from customer requirements, marketplace, engineering, laboratory research, etc. As new products are intended to satisfy the needs of the consumers, each design solution is influenced by various factors that must be considered when making decisions about colors, shapes, and materials. Factors like socio-economic, cultural, and technological development influence market demands, consequently playing an important role in design decisions alongside ergonomics, aesthetics, ecological demands, etc. Keeping in mind the aforementioned factors and their complexity, the importance of gathering quality information and conducting proper research is obvious. Valuable information for idea generation comes from primary and secondary data sources like focus groups, in-depth interviews, observations, product analysis, conducting workshops, demographic data, etc. The overview of data sources is shown in Table 1.

Outcomes of the research stage are textual and statistical reports for managerial decisions important for formulating the product requirements.

Table 1 Data sources defining design problem and requirements

Data (Facts and figures about design problem and requirements)	
Secondary data Facts and figures known and recorded prior to the project	*Primary data* Facts and figures newly collected specifically for the project
Internal data Financial statements, research reports, files, customer letters, sales call reports, and customer lists	*Observational data* Mechanical and electronic approaches, Personal approaches
External data Census reports, trade association studies and magazines, business periodicals, and Internet-based reports	*Questionnaire data* In-depth interviews and focus groups mail, online, telephone, and personal surveys

A detailed design brief resulting from the research conclusions is the first step in successful concept design. Concept design is usually the most unrestricted stage of the idea generation process, allowing for the utilization of numerous suitable techniques. Some of the most used are brainstorming, mind mapping, storyboarding, role-playing, forced relationships, C-sketch (collaborative sketching), SCAMPER, 6-3-5, etc. Each of these methods has its advantages and disadvantages [3–6]. Generally, techniques like 6-3-5 and C-Sketch produce a larger number of ideas as participants have insight into a subset of all the ideas. On the other hand, methods such as brainstorming and storyboarding allow for faster idea development where participants can view all the ideas the whole team has generated during the idea generation process [3]. This stage results in numerous ideas captured as sketches or crude models intended only to clarify the idea.

Idea refinement is a process of selection and detailed definition of an idea. This phase in a product design process pushes the idea further towards the realization of the product, still offering vast possibilities and requiring consideration of many variations. It is the final stage of idea generation, after which the design team should have a clear idea about the product and potential market response to the product. The design team at this time has the scope of the functional, technical, market, and business aspects of the product concept developed in the previous stages. Early mockups are presented in order to get the market feedback and test if the concept is good and if it satisfies customer demands and expectations.

The complexity of the idea generation process emphasizes the importance of optimization, which affects time resources and general ability to develop concepts that have the potential for achieving aesthetic, ergonomic, performance, buildability, and other requirements, as well as ensuring the consumer's satisfaction. Digital transformation of the idea generation process in the past decades and constant adaption of the emerging technologies offer tools for the fast and successful journey through this process.

2 Transformation of the Market Research Methodologies by Digital Technology

Traditional data gathering methods still have their place in the research methodologies, but the availability of new technologies is rapidly changing trends and practices in this field. The introduction of the Internet changed the way people communicate by offering digital alternatives to all forms of human communication, from written messages to face-to-face interaction. In the last three decades, digital communication went from the basic text to video communication as we know it today, with virtual environments fast approaching. In one minute, 168 million e-mails are sent, 694,445 Google searches are performed, 695,000 Facebook status updates are posted, 370,000

Skype calls are made, 98,000 tweets on Twitter are made, etc. [7]. Keeping in mind the current consumption patterns, products life cycle, and development period becoming shorter and shorter, this evolution of the communication and new opportunities for gathering information for product development had to be used. Internet research developed very dynamically in the second half of the 90s [8]. At first, surveys were digitized, offering time and money-efficient methods of data gathering. Like other research methods, online surveys have their benefits and drawbacks, depending on the specifics of the project. Some of the benefits are quick response, low cost, and easy fielding, but at the same time, there are some drawbacks, such as low response rates and claims that samples do not adequately represent populations. Online surveys show their superiority when: there is a need for a large and geographically widely distributed sample size; the questionary can benefit from the use of multimedia or interactive features in the questionary; the information is not sensitive for the participant and if the targets are Internet users [9].

Today social networks have become an everyday part of human life. They are rich in all types of content (video-audio, text, and images) and linked data, forming a vast source of various complex and heterogeneous data. Traditional analysis tools are inefficient in handling such data. Therefore, many data mining and artificial intelligence techniques exist and are still being developed to extract useful patterns and knowledge from this data. Data mining could be explained as an umbrella term covering and combining statistics, artificial intelligence, and machine learning to search and analyze data in order to find implicit, potentially useful information, eliminate the randomness, and uncover previously unknown patterns by selecting, exploring, and modeling large amounts of data from a large database [10].

The typical flow of the data-driven analysis of the consumer product requirements is shown in Fig. 1.

There are different tools available, but their basic functions are similar [11]:

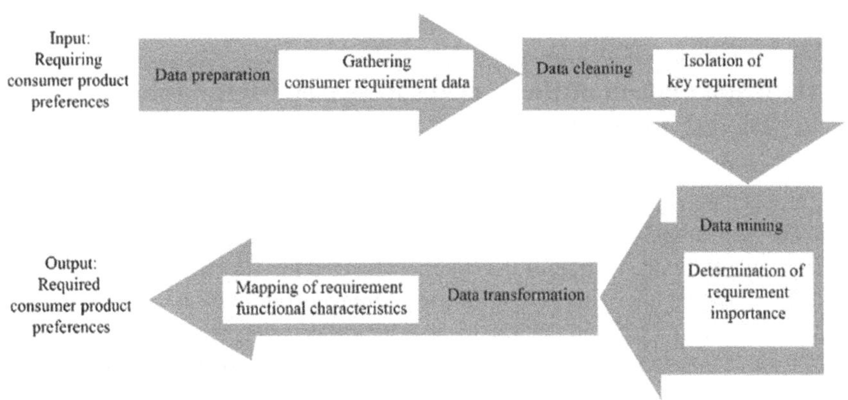

Fig. 1 Data-driven consumer product requirement analysis flow

- pattern discovery,
- trends detection,
- data dimensionality reduction and
- visualization.

Some of the most widely used data-mining algorithms are [12]:

- decision-tree algorithms,
- decision-rule algorithms,
- Bayesian algorithms,
- neural networks,
- clustering and
- regression.

In a digital environment saturated with data, the fundamental challenge is not collecting the data, but rather comprehensively organizing unstructured data and extracting useful conclusions.

Product intelligence is an emerging field of collecting and analyzing data about a product's performance with users, creating a positive feedback loop helping the design team to innovate fast, accurately and efficiently satisfying user requirements. Apple is one of the first to implement product intelligence. They gather data about how consumers use their device directly from the device, and use that information to make changes and keep improving products to constantly be competitive in the smartphone industry. Furthermore, the introduction of smart products in all segments of consumer goods production, and the spreading of the Internet of Things concept help offer direct communication between products and manufacturers, which can contribute to product improvement resulting in a good customer experience.

3 Artificial Intelligence in the Concept Design and Idea Generation

Generating many design idea variations and choosing the best of them according to the predetermined performance criteria is a goal of every design process. A number of the idea variations greatly depend on the resources (human resources, time, money, etc.). Limitations in the resources usually result in a finite number of variations and a relatively small number compared to the number of viable variations that should be considered. With the introduction of artificial intelligence (AI) in idea generation, process designers can create design solutions faster and cheaper due to the speed at which AI can analyze vast amounts of data and suggest design variations. The designer can make the final decision or opt for the performance criteria broadening. The use of artificial intelligence in providing design has proven great benefits, especially in structural optimization tasks. Using the Variational Autoencoders (VAEs) model to create AI-generated design ideas results in optimization and a great diversity of design outputs. These AI models can conduct the learning process on an artificial dataset to generate design ideas [13]. Generative design is not constricted by previous experience and often results in uncommon solutions that do not fall within the typical set of shapes or configurations. As a result, it has been first exploited mainly to encourage divergent thinking and creativity [14]. Philippe Starck's A.I. chair, is claimed to be the first production chair designed using artificial intelligence [15].

Engineering data science propelled by the increased availability of computing power is combining the potentials of AI with the robust simulation, testing, and field data sets in all segments of product development. This helps in removing strict boundaries between concept design, design development, engineering, and technology by applying the same physics-based tools used for verification in each product development stage. Using predefined constraints, AI-supported product development systems enable faster convergence of design by ranking design solutions and rejecting those with lower potential early in the development process. There are numerous examples of successful use of AI in product development of well-known manufacturers in the automobile industry, electronic industry, etc., documenting advantages in simplifying complex systems, deploying early detection for fast action, increasing precision in processes [16].

The product development process is at the brink of automation by AI technology implementation. The near future will enable designers to set design goals, prices, and choose materials, implement that information into the AI algorithms and get numerous design solutions in short time. These proposed solutions will be based on the data collected and processed through cloud computing and big data technologies. Product development in conjunction with changes in production technologies and adoption of additive manufacturing could change the whole design and manufacturing process, offering products highly customized to the consumer needs [17]. The proposed concept is presented in Fig. 2.

As the design process is a collaborative activity, real-time collaboration over the Internet can support the process of idea generation. Brainstorming is one of the best-known and most used collective idea generation methods. It was modernized by the introduction of the Internet and branded as Electronic Brainstorming or Brainneting. Effective Brainstorming needs a moderator. Artificial intelligence functions like natural language processing, machine learning, and reasoning could help create comprehensive virtual moderators. This enables multiple participants to share their

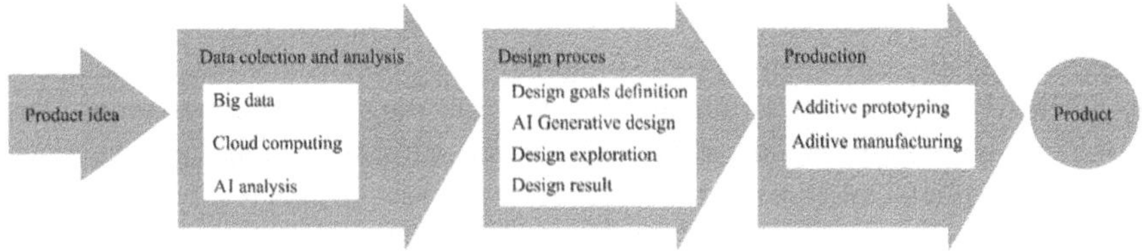

Fig. 2 Concept of automated design and manufacturing process

ideas without having to wait for their turn, as in face-to-face Brainstorming. Another benefit is overcoming the problem of evaluation apprehension, as participants are communicating with the computer and not with the other person [18].

AI is expected to play an important role in the future of product design. Numerous activities of the design process have already been affected. Although this technology is rapidly developing and being more and more implemented, total replacement of humans in the creative processes still seems impossible in the foreseeable future.

4 Design Process Supported by the Virtual and Augmented Reality Technologies

Visualization in idea generation is crucial. Traditionally sketches are used to capture and present the idea in a fast and loose manner. The goal is to present as much information as possible in as little time as possible. Transition to computer technology resulted in a similar two-dimensional digital sketch, as three-dimensional modeling takes time and effort, which is not affordable for the numerous idea capturing. With the introduction of virtual environment (VR) and tolls for VR-human interaction (headsets and controllers) possibility of producing fast three-dimensional representations of a design idea became possible. Virtual reality produces an entirely computer-generated simulation of an alternate world. These immersive simulations can create a life-like visual perception of the material objects using headsets, gloves, and sensors. Although still in its infancy, this technology captures the interest and imagination of designers, especially in the early creative design phases [19, 20]. This technology adoption in early design phases is also noticeable in products that involve an emotional dimension beyond functional aspects, virtual assemblies, and prototypes [20]. Although design verification is a very important step, it can prove to be a significant problem when design and managerial teams are in separate locations, especially in the automobile industry, considering the size of the product. Therefore, VR has attracted considerable attention. Research into the usability of this technology verified that with regard to the perception difference between actual cars and VR cars, there is no significant difference regarding the exterior of the vehicle, but when the objective is to evaluate the comfort of the interior space of a car, using VR is not recommended [21].

Augmented reality (AR) is an interactive experience where the real-world environment is augmented by computer-generated perceptual information such as sound and images. This technology offers possibilities for design presentation and collaborative design, with researchers reporting the following advantages: support for direct manipulation in part creation and modification; easy evaluation of modifications in real-time.

5 Conclusions

The design process at its early stages, namely idea generation, uses similar steps established before introducing digital technologies. However, each individual step suffered drastic changes through time under the influence of ever-evolving digital technologies. Results of research into consumer needs as a driving factor of every product development process have gotten extremely precise in identifying each of the product's characteristics. The introduction of the possibility of big data processing, product intelligence, and emerging of the Internet of Things enables further and easier refinement of the products to increase user satisfaction. However, the results of this process still greatly depend on the quality of ideas for solving identified problems and addressing opportunities.

Support of artificial intelligence in all aspects of idea generation is notable and welcomed by industry leaders, with great expectations of AI implementation in the future. VR and AR technologies have already found their place in the early design stages. With the improvement of their capabilities and the dissolving of the boundaries between the real and virtual world, their application is expected in all product development stages.

References

1. Cooper RG (2017) Winning at new products: creating value through innovation. Basic Books, New York
2. Cooper RG, Edgett SJ, Kleinschmid EJ (2004) Benchmarking best NPD practices-1: culture, climate, teams and senior management's role. Res-Tech Manag 47:31–43
3. Linsey JS, Green MG, Murphy JT, Wood KL, Markman AB (2005) "Collaborating to success": an experimental study of group idea generation techniques. Proceedings of IDETC/CIE 2005:277–290
4. Vogel T (2014) Breakthrough thinking: a guide to creative thinking and idea generation. Simon & Schuster, New York
5. Greiner D (2017) The basics of idea generation. Productivity Press, New York
6. Smith G (1998) Idea-generation techniques: a formulary of active ingredients. J Creat Behav 32:107–134
7. Panda M, Hassanien AE, Abraham A (2019) Big data analytics: a social network approach. CRC Press, Boca Raton
8. Ilieva J, Baron S, Healey N (2002) Online surveys in marketing research. Int J Mark Res 44:361–376
9. Sue VM, Ritter LA (2012) Conducting online surveys. Sage, Thousand Oaks
10. Huang TS, Chang, CF (2005) The role of data mining in the product design and development process. In: Proceedings of CAIDCD'05, pp 198–203
11. Kusiak A, Smith M (2007) Data mining in design of products and production systems. Annu Rev Control 31:147–156
12. Witten IH, Frank E (2005) Data mining: practical machine learningtools and techniques. Elsevier, New York
13. Mirra G, Pugnale A (2021) Comparison between human-defined and AI-generated design spaces for the optimisation of shell structures. Structures 34:2950–2961

14. Buonamici F, Monica C, Furferi R, Volpe Y, Governi L (2020) Generative design: an explorative study. Comput-Aided Des Applic 18:144–155
15. Starck (2019) A.I. for KartelL by Starck powered by autodesk (Kartell). https://www.starck.com/a-i-for-kartell-by-starck-powered-by-autodesk-kartell-p3534. Accessed 20 Nov 2021
16. MIT Technology Review Insights (2021) Product design gets an AI makeover. https://wp.technologyreview.com/wp-content/uploads/2021/05/Product-design-gets-an-AI-makeover.pdf?_ga=2.223889435.16837537.1638350004-360022625.1638350004. Accessed 20 Nov 2021
17. Hyunjin C (2020) A study on the change of manufacturing design process due to the development of A.I design and 3D printing. In: IOP Conference Series: Materials Science and Engineering. 727. 012010.
18. Strohmann T, Siemon D, Robra-Bissantz Susanne (2017) brAInstorm: intelligent assistance in group idea generation. In: Designing the Digital Transformation, pp 457–461
19. Cecil J, Kanchanapiboon A (2007) Virtual engineering approaches in product and process design. Int J Adv Manuf Technol 31:846–856
20. Berni A, Borgianni Y (2020) Applications of virtual reality in engineering and product design: why, what, how. When and Where. Electronics 9:1064
21. Kato T (2019) Verification of perception difference between actual space and VR space in car design. Int J Interact Des Manuf 13:1233–1244

Index

P. Kyratsis et al. (eds.), *Advances in Product Design Engineering*, Management and Industrial Engineering,
https://doi.org/10.1007/978-3-030-98124-2

The manufacturer's authorised representative in the EU is Springer Nature Customer Service Centre GmbH, Europaplatz 3, 69115 Heidelberg, Germany. If you have any concerns regarding our products, please contact ProductSafety@springernature.com

Printed and bound by CPI Group (UK) Ltd, Croydon, CR0 4YY
30/09/2026
02225324-0006